AGRICULTURE ISSUES AND POLICIES

ZEA MAYS L.

CULTIVATION, AND USES

AGRICULTURE ISSUES AND POLICIES

Additional books and e-books in this series can be found
on Nova's website under the Series tab.

AGRICULTURE ISSUES AND POLICIES

ZEA MAYS L.

CULTIVATION, AND USES

SARAH DUNN
EDITOR

Copyright © 2021 by Nova Science Publishers, Inc.

All rights reserved. No part of this book may be reproduced, stored in a retrieval system or transmitted in any form or by any means: electronic, electrostatic, magnetic, tape, mechanical photocopying, recording or otherwise without the written permission of the Publisher.

We have partnered with Copyright Clearance Center to make it easy for you to obtain permissions to reuse content from this publication. Simply navigate to this publication's page on Nova's website and locate the "Get Permission" button below the title description. This button is linked directly to the title's permission page on copyright.com. Alternatively, you can visit copyright.com and search by title, ISBN, or ISSN.

For further questions about using the service on copyright.com, please contact:
Copyright Clearance Center
Phone: +1-(978) 750-8400   Fax: +1-(978) 750-4470   E-mail: info@copyright.com.

NOTICE TO THE READER

The Publisher has taken reasonable care in the preparation of this book, but makes no expressed or implied warranty of any kind and assumes no responsibility for any errors or omissions. No liability is assumed for incidental or consequential damages in connection with or arising out of information contained in this book. The Publisher shall not be liable for any special, consequential, or exemplary damages resulting, in whole or in part, from the readers' use of, or reliance upon, this material. Any parts of this book based on government reports are so indicated and copyright is claimed for those parts to the extent applicable to compilations of such works.

Independent verification should be sought for any data, advice or recommendations contained in this book. In addition, no responsibility is assumed by the Publisher for any injury and/or damage to persons or property arising from any methods, products, instructions, ideas or otherwise contained in this publication.

This publication is designed to provide accurate and authoritative information with regard to the subject matter covered herein. It is sold with the clear understanding that the Publisher is not engaged in rendering legal or any other professional services. If legal or any other expert assistance is required, the services of a competent person should be sought. FROM A DECLARATION OF PARTICIPANTS JOINTLY ADOPTED BY A COMMITTEE OF THE AMERICAN BAR ASSOCIATION AND A COMMITTEE OF PUBLISHERS.

Additional color graphics may be available in the e-book version of this book.

Library of Congress Cataloging-in-Publication Data

ISBN: 978-1-53619-181-3

Published by Nova Science Publishers, Inc. † New York

CONTENTS

Preface		vii
Chapter 1	Disclosure of The Potential Phytase-Producing Maize Endophytic Bacteria, as an Invisible Avail for *Zea mays* L. *Hafsan, Cut Muthiadin, Eka Sukmawaty, Nurhikmah and Yuniar Harviyanti*	1
Chapter 2	Maize (*Zea Mays* L.) Suitability for Wet Milling and Animal Nutrition in Relation to Physical and Chemical Quality Parameters *Marija Milašinović-Šeremešić, Olivera Đuragić, Milica Radosavljević and Ljubica Dokić*	51
Chapter 3	The Effect of Salicylic Acid in Maize Bioproductivity *C. J. Tucuch-Haas, G. Alcántar-González, L. Trejo-Téllez, H. Volke-Haller, Y. Salinas-Moreno, J. I. Tucuch-Haas, M. A. Dzib-Ek, S. Vergara-Yoisura and A. Larque-Saavedra*	83
Index		99

PREFACE

Zea mays L. is a potential producer of cereal crops and the dominant primary energy source of feed for monogastric animals, such as poultry. The first chapter in this book aims to determine the potential of phytase-producing endophytic bacteria, as an invisible avail for *Zea mays* L. High phytate levels in maize seeds is a problem encountered when used as raw material in poultry feed. The second chapter of this book focuses on the physical traits, chemical composition, and their relationship with wet-milling properties and nutritional quality parameters of maize hybrids of different maturity groups and various endosperm types (dent, semi-dent and flint). Finally, Mesoamerican cultures are generally regarded as advanced societies that, among other contributions to humanity, are known to have domesticated cultivated plants as *Zea mays*. Maize is one of the staple foods of the Mexican population and the practice of nixtamalization of maize seeds before Spanish conquest in 1521, is fundamental in the preparation of dough for tortillas. The last chapter examines the effect of salicyclic acid in maize bioproductivity.

Chapter 1 - *Zea mays* L. is a potential producer of cereal crops and the dominant primary energy source of feed for monogastric animals, such as poultry. The potential energy content of maize seeds, expressed as Metabolizable Energy (ME) is relatively high compared to other feed ingredients. However, they contain phytic acid, which acts in physiological functions (storage of main Phosphorus and cations). In Maize seeds, phosphorus is primarily stored as phytate (60-97%), and approximately 21-25% are found in the plant root. This study aims to determine the potential of phytase-producing endophytic bacteria, as an invisible avail for *Zea mays* L. High phytate levels in maize seeds is a problem encountered when used as raw material in poultry feed. The inability of poultry to produce phytase in hydrolyzing phytate feed in the digestive tract, reduces broiler digestibility of Phosphorus and other minerals banded by this element. This response is due to phytic acid effect, a potent chelator categorized as an antinutrient. Interestingly, it has the ability to bind to proteins and ions of several essential minerals, such as calcium, iron, zinc, magnesium, manganese, and copper, even forming complexes with digestive enzymes. A phytate problem-solving effort required the utilization of phytase enzyme from various sources, including those obtained from endophytic bacteria. The existence of phytic acid in phosphorus storage of maize plants, allows the presence of endophytic bacteria that utilize this acid as a metabolic substrate. Phytic acid is used as a source of Phosphorus for metabolic needs, and also engage in mutualism interaction with maize plants, since its life cycle does not have a detrimental impact on the host plant. By producing extracellular enzymes, phytic acid is hydrolysed by phytase, produced by endophytic bacteria. It is known that endophytic bacteria play a role in increasing plant growth and yield, suppressing contaminant pathogens, dissolving phosphates, or contributing nitrogen. Endophytic bacteria is one of the unique groups of organisms that have natural habitats in plant tissue, both in the root, leaves, stems, and seeds, and are fascinating to explore. Various secondary metabolites have been produced and studied, both as antibiotic, antiviral, anticancer, antioxidant, anti-insecticidal, antidiabetic, and anti-immunosuppressive compounds. The ability to produce phytase, applied in improving the quality of poultry feed was explored. The result found and identified four types of potential phytase-producing endophytic bacteria from the Maize plant, namely *Burkholderia* strain HF.7, *Enterobacter cloacae, E. ludwigii,* and *Pantoea stewartii*. Sequentially, each was isolated from the roots, stems, leaves, and seeds of the maize plants.

Chapter 2 - Maize (*Zea Mays* L.), being very competitive as a high carbohydrate yielding plant, represents the most important raw material for the commercial starch production (wet-milling industry) and a major energy feed ingredient in almost all animal diets. Maize kernel and whole plant varies in compositional traits and digestibility due to genetics and numerous environmental factors. The focus of this study is on the physical traits, chemical composition, and their relationship with wet-milling properties and nutritional quality parameters of maize hybrids of different maturity groups and various endosperm types (dent, semi-dent and flint). The selected hybrids were grown under the semiarid conditions with the application of the same cropping practices at the experimental field of the Maize Research Institute in Zemun Polje, Belgrade, Serbia. Furthermore, the aim was to characterize suitability of the maize hybrids for wet-milling and animal nutrition. Through many years of the authors' research of maize quality and utilization, the obtained results showed significant relationship between quality parameters and contributed in classifying their importance and relevance. Accordingly, obtained relationships enabled the estimation and prediction of maize quality (utility value) for a particular purpose. A significant negative correlation was found between kernel protein content and portion of soft endosperm as well as a significant positive correlation between kernel protein content and two physical parameters, milling response and density. Among the chemical composition parameters only starch content significantly affected the starch yield and negatively affected the gluten yield. Physical parameters of the kernel such as test weight, kernel density and hardness significantly affected starch yield and recovery. Hybrids with a lower test weight and density and a greater proportion of soft endosperm fraction had a higher yield, recovery and purity of starch. A significant negative correlation was determined between the NDF (Neutral Detergent Fibres) content and the whole plant dry matter digestibility (IVDMD), as well as, between the hemicellulose content and the digestibility. Based on the results, it can be concluded that a maize hybrid, intended for a high value animal feed, should have a low content in ADL/NDF (ADL-lignin NDF ratio, %), because it negatively influences whole plant digestibility. Furthermore, a very significant positive correlation was also found among all assayed lignocellulose fibers components. The study demonstrates the importance of evaluating both the chemical and physical quality parameters of the maize kernels due to screen and estimate the suitability (utility value) of maize kernel for wet milling and animal nutrition. Understanding how maize plant cell wall constituents affect IVDMD is an important goal of future breeding research programs in order to improve forage utilization in animal feeding. The results obtained in the authors' studies indicate that genetic differences in the long-term of maize hybrids breeding programs development can lead to providing farmers and industry with hybrids of good quality, desirable properties, and acceptable yield under the variable climatic conditions and with a lower cost.

Chapter 3 - Mesoamerican cultures are generally regarded as advanced societies that, among other contributions to humanity, are known to have domesticated cultivated plants as *Zea mays*. Maize is one of the staple foods of the Mexican population and the practice of nixtamalization of maize seeds before Spanish conquest in 1521, is fundamental in the preparation of dough for tortillas. The authors have shown that applications of low concentrations of salicylic acid (SA) in plant seedling shoots or in evergreen trees significantly increase growth, development, and productivity. In order to assess the effect of spraying, low concentrations (SA) in maize seedling in development conduct experiments in growth rooms that have shown that 1 μm of SA significantly increased root length by 30.6% and 0.1 M of SA 24.7% compared to control. This concentration also significantly increased the total fresh biomass of seedlings. In other experiments the results have shown that (SA) significantly increase the length, weight and dry weight of roots, stems, leaves and yield of this species, as well nitrogen (N), phosphorus (P) and potassium (K) levels in the different organs of plants at harvest time. Copper, zinc, manganese, iron, boron, calcium, and magnesium were also increased in most tissues by the effect of SA. It is proposed that the positive effect of SA of increasing root size promotes the absorption and accumulation of macro and micronutrients and contributes to seed production.

Chapter 1

DISCLOSURE OF THE POTENTIAL PHYTASE-PRODUCING MAIZE ENDOPHYTIC BACTERIA, AS AN INVISIBLE AVAIL FOR *ZEA MAYS* L.

Hafsan[*]*, Cut Muthiadin, Eka Sukmawaty, Nurhikmah and Yuniar Harviyanti*
Department of Biology, Universitas Islam Negeri Alauddin,
South Sulawesi, Indonesia

ABSTRACT

Zea mays L. is a potential producer of cereal crops and the dominant primary energy source of feed for monogastric animals, such as poultry. The potential energy content of maize seeds, expressed as Metabolizable Energy (ME) is relatively high compared to other feed ingredients. However, they contain phytic acid, which acts in physiological functions

---

[*] Corresponding Author's E-mail: hafsan.bio@uin-alauddin.ac.id.

(storage of main Phosphorus and cations). In Maize seeds, phosphorus is primarily stored as phytate (60-97%), and approximately 21-25% are found in the plant root.

This study aims to determine the potential of phytase-producing endophytic bacteria, as an invisible avail for *Zea mays* L. High phytate levels in maize seeds is a problem encountered when used as raw material in poultry feed. The inability of poultry to produce phytase in hydrolyzing phytate feed in the digestive tract, reduces broiler digestibility of Phosphorus and other minerals banded by this element. This response is due to phytic acid effect, a potent chelator categorized as an antinutrient. Interestingly, it has the ability to bind to proteins and ions of several essential minerals, such as calcium, iron, zinc, magnesium, manganese, and copper, even forming complexes with digestive enzymes. A phytate problem-solving effort required the utilization of phytase enzyme from various sources, including those obtained from endophytic bacteria.

The existence of phytic acid in phosphorus storage of maize plants, allows the presence of endophytic bacteria that utilize this acid as a metabolic substrate. Phytic acid is used as a source of Phosphorus for metabolic needs, and also engage in mutualism interaction with maize plants, since its life cycle does not have a detrimental impact on the host plant. By producing extracellular enzymes, phytic acid is hydrolysed by phytase, produced by endophytic bacteria. It is known that endophytic bacteria play a role in increasing plant growth and yield, suppressing contaminant pathogens, dissolving phosphates, or contributing nitrogen.

Endophytic bacteria is one of the unique groups of organisms that have natural habitats in plant tissue, both in the root, leaves, stems, and seeds, and are fascinating to explore. Various secondary metabolites have been produced and studied, both as antibiotic, antiviral, anticancer, antioxidant, anti-insecticidal, antidiabetic, and anti-immunosuppressive compounds. The ability to produce phytase, applied in improving the quality of poultry feed was explored. The result found and identified four types of potential phytase-producing endophytic bacteria from the Maize plant, namely *Burkholderia* strain HF.7, *Enterobacter cloacae*, *E. ludwigii,* and *Pantoea stewartii*. Sequentially, each was isolated from the roots, stems, leaves, and seeds of the maize plants.

Keywords: endophyte, phytase, phytic acid, maize, *Zea mays* L., *Burkholderia*

1. INTRODUCTION

Maize (*Zea mays* L.) is an annual crop commonly cultivated by residents of various communities, and is the world's largest food source after rice and wheat. It is a cereal that has strategic economic value, and has the opportunity to be expanded because of its position as the primary source of carbohydrates and protein for humans. Maize is also widely used for various purposes, such as feed ingredients for farm animals, humans, and also an essential component in the production of other products including ethanol fuel, adhesives (glue), cosmetics, hand soap, etc. Its plant biomass in the form of stems and leaves is also used for green manure and animal feed. The listed world maize consumption is found to be increasing on a daily basis, including meeting the needs in feed production. In 2020, its global demand is found to increase by 45% [1].

Maize containing about 72% starch, 10% protein, and 4% fat provides relatively high metabolic energy (EM), reaching 365 Kcal/100 g [2]. This has led to the use of seed maize as the present primary energy source in feed, which has not been replaced. For monogastric farm animals, such as poultry, their primary potential source of energy for metabolism are maize seeds, compared to other feed ingredients [3, 4]. This is due to the limitations of poultry using different crude fibre from a polygastric animal. However, maize seeds contain phytic acid that plays a role in their physiological functions (primary phosphorus storage and cations). During dormancy and germination, the phytic acid in maize plays a role in protecting oxidative damage in the storage process. Acting as the central reserve, phytic acid is 85% of the total Phosphorus (P) in cereals and legumes [5-7].

The high level of phytate in maize seeds is a crucial problem in its use as the main ingredient in poultry feed. This is due to the nature of the acid as a potent chelator categorized as anti-nutrition. Also, it has the ability to bind proteins, ions, and some essential minerals, such as calcium, iron, zinc, magnesium, manganese, and copper [8]. The presence of phytate compounds reduce the digestibility of Phosphorus, protein, and other minerals found in feeds, because they are unhydrolysable in the digestive tract [9].

Phosphorus and calcium are crucial elements for all animals, including poultry, needed for bone mineralization, immunity, fertility, and general growth. It is essential to maintain the availability of Ca and P that are digested by broiler to support metabolism [10]. The efforts to increase the efficiency of Phosphorus and other vital minerals bind by phytic acid, reduce its negative effect on nutrient utilization. However, the bonds are broken by hydrolysis process [9]. Poultry of monogastric animals have limitations in producing phytase in their digestive tract [11, 12]. To meet the phosphorus needs of poultry, it is usually necessary to add inorganic Phosphorus, such as dicalcium or monocalcium phosphate to the feed. Consequently, this causes an increase in the amount of Phosphorus wasted with faeces into the environment [13, 14], and simultaneously have implications in ecological damage, such as leading to the occurrence of water eutrophication [15, 16].

The existence of phytic acid as a storage form of Phosphorus in maize plants allows the presence of endophytic bacteria. Phytic acid is used as a source of Phosphorus for the metabolic needs of bacteria, in addition to its symbiotic interaction with the host plant [17]. Also, it is known that the acid life cycle does not have a detrimental impact on the host plant [18]. By producing extracellular enzymes, phytic acid is hydrolysed by the phytase they produce [19-21]. Moreover, it is known that endophytic bacteria play roles in increasing plant growth and yield, suppressing pathogenic contaminants, dissolving phosphates, or contributing nitrogen [22-24].

Endophytic bacteria are a unique group of organism having a natural habitat in plant tissue, such as root, leaves, stems, and seeds, and are fascinating to explore [18]. Various secondary metabolites have been produced and researched, both as antibiotic, antiviral, anticancer, antioxidant, anti-insecticide, antidiabetic, and anti-immunosuppressive compounds [23-25]. The ability to produce phytase has not been widely reported, especially their application in improving the quality of poultry feed. However, prospecting endophytic bacteria as a potential producer of phytase maize crop has been explored.

Maize plant endophytic bacteria have the potential to produce phytase through substrate induction, to be used as an additive for poultry feed. The phytase produced supports the prospect efforts in improving poultry feed quality, by optimizing the release of minerals and protein in the meal [12, 20, 21, 26, 27). This increases productivity because it reduces feed production costs, and the use of inorganic Phosphorus which is relatively expensive [10, 15, 16]. Furthermore, it also improves digestibility and performance of broilers because it is supported by the adequacy and absorption of feed nutrients [10, 14, 26, 28, 29], which leads to the creation of environmentally friendly animal husbandry [13, 30].

Phytase is an enzyme of phosphomonoesterase forming a monomeric protein [31], which hydrolyse phytate to inorganic orthophosphate, Myo-inositol, monophosphate, free protein, and other minerals bound to the myo-inositol group [32, 33]. The working principle of phytase in nutrient utilization is by breaking the bonds of phytate compounds in minerals and proteins, to be maximally utilized in the process of metabolism and biosynthesis [19, 29, 34].

Several studies have been conducted to determine the various sources of phytase and their effect on the availability of Phosphorus in monogastric animal feed. [32] The results showed that the use of inorganic Phosphorus is minimized, and it is estimated that 10 kg of calcium phosphate is replaced with only 0.25 kg phytase [10, 28]. The use of inorganic Phosphorus tends to be expensive, and reducing it undoubtedly decreases feed costs [15, 35, 36]. Also, the use of inorganic Phosphorus reduces its amount released through faeces, therefore, decreasing environmental pollution [37].

Several other studies have shown that phytate supplementation in feed increase the use of Phosphorus which binds to phytates. [38] It was also observed that 500 U/kg of Natuphos® phytase enzyme supplementation in broiler chicken feed containing low available P (0.22%), was able to improve performance and increase the use of P, Ca, Mg, and Zn. Phytase application also increase the bioavailability of protein and minerals through phytate hydrolysis in the digestive tract or during the process of making feed [39, 40]. Other studies have reported that the addition of phytase in broiler feed improve the bioavailability of amino acids, arginine, and other minerals [41]. In another study, the effect of phytase treatment on ileal digestibility of amino acids was found to have a significant impact on wheat-based feed. Individually, phytase increases the digestibility of ileum arginine, histidine, isoleucine, leucine, lysine, methionine, phenylalanine, threonine, aspartic acid, glutamic acid, glycine, proline, and serine from 2.5% to 12.8% [34].

Phytase is obtained from various sources, presently, many have been collected from plants, fungi, bacteria, and the rumen of ruminant animals [11]. Bacteria as a source of enzymes, have more value compared to those isolated from animals and plants. Among others, because bacterial cells are relatively more comfortable and faster to grow, their scale of cell production is more accessible for more excellent yield through the regulation of growth conditions and genetic engineering. Besides, the conditions observed during their production are not limited by the change of seasons, as well as a more uniform quality [19, 42]. By this fact, it is critical to focus on phytase-producing bacteria in the search for an excellent source of enzymes. This is related to variations in the characteristics of an enzyme produced by a different source, such as substrate specificity, catalytic efficiency, as well as other physiological properties. Presently, some enzyme from a strain of bacteria has been isolated, cloned, or expressed as phytase from microbes, namely *Escherichia coli, Bacillus sp., B. amyloliquefaciens, B. licheniformis, B. coagulans, B. stearothermophillus, Geobacillus, Lactobacillus amylovorus, Burkholderia, Enterobacter cloacae, E. ludwigii, Pantoea stewartii, Selenomonas ruminantium, Klebsiella pneumonia, K. oxitoca, K. aerogenes,* and *K. terrigena* [19, 20, 35, 43, 44-47].

2. METHODS

2.1. Isolation, Screening, and Characterization of Phytase-Producing Endophytic Bacteria from Maize Plants

In order to obtain and determine the characteristics of maize plant endophytic bacteria which produces phytase. The process begins with the isolation of bacteria from the maize plant organ including aseptic preparation of root, stem, and seed samples. The bacterial selection was based on isolates that had the highest phytatic index (PI), on selective media for phytase from each of the four maize plant organs. The selected isolates showed the highest ability to hydrolyse phytate based on PI, namely the ratio between the diameter of the clear zone around the growing colony [48-53]. The isolates were then characterized based on cell and colony morphology, as well as Gram characteristics before identification using a molecular approach.

2.1.1. Media Preparation

The media used consists of isolation, selective, and phytase production media. The isolation medium used was Luria Bertani with a composition per litre: 10 g peptone, 5 g yeast extract, and 10 g NaCl. The selective medium used was Phytase Selective Media (PSM) with a composition per litre: 15 g glucose, 5 g (NH4) 2SO4, 0.1 g NaCl, 0.5 g KCl, 0.01 g FeSO, 0.1 g MgSO4 .7H2O, 0.1 g CaCl2.2H2O, 0.01 g MnSO4, and 4 g of Na-Phytate. The phytase production media used was Phytase Production Media (PPM) with ingredients per litre: 15 g glucose, 5 g Na-phytate, 5 g NH4SO4, 0.5 g KCl, 0.5 g MgSO4.7H2O, 0.1 g NaCl, 0.01 g CaCl2.2H2O, 0.01 g FeSO4.7H2O, and 0.01 g MnSO4.H2O. This process was carried out by dissolving all the materials that have been weighed carefully in 1000 mL of sterile distilled water in a beaker glass, then homogenized using a hot plate magnetic stirrer until all the ingredients dissolve, and adjusting the pH of the media, namely pH 6. Then sterilization was carried out by autoclaving at 121°C for 15 mins at of 2 atm.

2.1.2. Sample Preparation

The samples used were four organs from a 110-day old maize specimen consisting of roots, stems, leaves, and seeds. Each organ was separated, then cleaned with running water, and cut into small pieces. The surface was sterilized by immersing in sodium hypochlorite for 2 mins, 70% ethanol for 2 mins, and 96% ethanol for 2 mins. Each sample was rinsed with sterile distilled water twice, then crushed aseptically using a mortal and pastle.

2.1.3. Isolation of Endophytic Bacteria from Maize Plants

This began with 10 g of samples from each prepared organ cultivated on 90 mL LB medium. As a control for the sterility of the sample surface, distilled water was also cultivated. When the media did not show any bacterial growth, it was ascertained that the species obtained were endophytic. The liquid culture was then incubated on a shaker incubator for 1 x 24 hours at a speed of 100 rpm (rotation per minute). The culture was serially diluted up to 10^{-8} dilution to avoid too dense growth on agar plate culture. The dilution of 10^{-6}, 10^{-7}, and 10^{-8} were inoculated on solid LB media with dispersive method, then incubated at 28°C for 1 x 24 hours. Colonies that grew and manifested diverse characteristics indicated different bacteria. The various colonies are then purified by scratching on the same media to obtain a type of bacteria that does not mix with other species/strains. Purification was carried out by taking one loop of separate bacterial colonies, and scratching on the media for a new sterile similar plate to be incubated at 28°C for 1 x 24 hours. From the growing colonies, re-scratching was carried out on a new solid medium to obtain genuinely pure isolates which were marked as single colonies formed at the end of the streak. The pure isolates were stored in the medium in slant storage at 4°C.

2.1.4. Screening for Phytase-Producing Endophytic Bacteria

Each isolate was inoculated from the culture stock into the selective media for phytase-producing bacteria, namely agar plate PSM using a bottle and simultaneously incubated at 28°C for 1 x 24 hours. Bacterial isolates that produce phytase showed a clear zone around their colonies. The isolates with the highest PI (as superior isolates) were selected and stored at 4oC for subsequent purposes.

2.1.5. Characterization of Bacterial Isolates

Isolates with the highest phytatic index were characterized by macroscopic, microscopic, and Gram characteristic observations. The macroscopic observations include remarks of size, pigmentation, shape, elevation, surface, and colony margins. The microscopic observations were carried out using Gram staining to observe the shape of cells and characteristics of the isolates.

2.2. Identification of Phytase-Producing Maize Plant Endophytic Bacteria Using a Molecular Approach

Identification of selected maize plant endophytic bacterial isolates that produces phytase was carried out to the species level using a molecular approach. This method was a validation of phenotypic identification that had been carried out based on the morphological, physiological, and biochemical characteristics of the selected isolates. This is necessary and gives many similarities in the biochemical and physiological properties possessed by different bacteria. The selected isolates were identified molecularly by analysis of the 16s rRNA gene [54]. The stages of identification of phytase-producing Maize plant endophytic bacteria using a molecular approach were as follows:

2.2.1. Rejuvenation of Phytase Enzyme-Producing Endophytic Bacterial Isolates

A pure culture collection of selected endophytic bacterial isolates was purified up to eleven times on Luria Bertani media by repeating streaking and incubating for 1 x 24 hours at 28°C. The cultures that grew in the final purification were prepared as stock on Luria Bertani media to be slanted for the DNA extraction stage.

2.2.2. DNA Extraction

2.2.1.1. Sample Preparation

One loop of the bacterial sample was placed in a sterile 1.5 mL microcentrifuge tube containing 200 µL Gram (+) buffer, which had been added with lysozyme. Then homogenize by pipetting and incubated at 28° C for 30 mins. The tube was vortexed, then added with 20 µL proteinase K and 200 µL Gram (-) buffer, vortexed and incubated again at 60° C for 10 mins. At every 3 mins the tube was turned back and forth to maintain homogeneity.

2.2.1.2. Cell Lysis

First, 200 µL of BG (Buffer Geneaid) was added to the sample then vortexed and incubated again at 50° C for 10 mins, then turning the tube at every 3 mins.

2.2.1.3. DNA Binding

Add 200 µL of absolute ethanol 96% and vortex for 10 seconds. The whole mixture was transferred to a spin column in a collection tube then centrifuged at a speed of 13,100 rpm for 2 mins. The collection tube under the spin column was discarded and replaced with new.

2.2.1.4. Washing

First, 400 µL W1 buffer (Geneaid) was added, then centrifuged at 13,100 rpm for 30 secs, after this, the liquid in the collection tube was discarded. Adding 600 µL of wash buffer and centrifuged again, also, the liquid in the collection tube was discarded and centrifuged again for 30 secs, then dumped the liquid in the collection tube. The collection tube that was under the spin column was removed and replaced with new. Then, it was again centrifuged at 13,100 rpm for 3 mins until the matrix column was dry.

2.2.1.5. Elution

100 µL of elution buffer (Geneaid) was added, and left standing for 3 mins then centrifuged at the same speed for 1 min. The liquid containing DNA stored in the microcentrifuge tube was kept at 4° C to be used as a template in the DNA amplification process with Polymerase Chain Reaction (PCR).

2.2.3. DNA Amplification by PCR

The polymeration chain reaction stage is an enzymatic synthesis process to multiply a specific nucleotide sequence in vitro (in a PCR tube). The method includes three stages, namely denaturation, annealing, and extension. This procedure is performed on DNA samples that have been isolated and extracted at a previous step. This stage was conducted by inserting the PCR mix into the its tube. The composition of the 25 µL PCR mix was: 9.5 µLddh2O, 12.5 µL PCR master mix, 5 µL 63F (Forward primer), 5 µL 1387R (reverse primer), and 2 µL of template DNA. The total PCR mix was 25 µL for each sample, then entered in a PCR (DNA thermal cycler) machine to amplify the DNA of the bacterial isolates. The use of this machine began with the pre-denaturation stage at 94°C for 2 mins, denaturation at 94°C for 1 min, annealing at 58°C for 45 secs, extension at 72° C for 90 secs and 30 cycles, followed by a final extension at 72° C for 5 mins, and the last was held at 4°C.

2.2.4. Electrophoresis

The electrophoresis process began with the manufacture of agarose gel, which was carried out by dissolving 2 g of agarose (2%) in 100 mL 10 X Tris borate EDTA. Then heat it to a boil and dissolve using a hot plate and stirrer. Then added with 1 mL ethidium bromide (0.2 mg/mL) and placed into the printer gel that has been fitted with a comb. After the agarose solidified, it was placed into an electrophoresis tank containing 0.5% TBE solution. A total of 5 µL of the amplified DNA sample was added, and to determine the size of the PCR amplification product, a 100 bp marker was inserted in the first well, followed by the amplified DNA sample in the second well and so on. As a ballast, 2 µL of loading dye was added for each DNA amplified sample, then homogenized by pipetting. Furthermore, the electrode was connected to the power supply then turn it on for 60 mins with 100 volts. After that, the electrophoresis tool was turned off, and the gel was taken and transferred into a gel doc tool, then the results was read on a computer.

2.2.5. Sequencing

The PCR processed samples and the 63F primers were sent to 1st BASE Malaysia for sequencing. The result was nucleotide sequences ± 1300 bp long. And were analyzed using the Basic Local Alignment Search Tool (BLAST) programme from the National Center for Biotechnology Information (NCBI) on the website (https//www.ncbi.nlm.nih.gov), to match the species data in the gene bank. The identities used were in the 80-100% range. Most similar Gene Bank sequences were characterized by the same Max and Total Score, Query Coverage close to 100%, E-value close to 0, and Max Ident close to 100%. To determine the level of kinship between species, the sequence alignment was carried out using the Clustal W. programme, then the construction of phylogenetic trees using the neighbour-joining method and the Molecular Evolutionary Genetics Analysis (MEGA) 5 programme.

2.3. Production and Optimization of Phytase Activity of Endophytic Bacteria from Maize Plants

This stage aims to obtain phytase from selected endophytic bacteria, and to determine the optimum temperature and pH of the resulting phytase activity. The mechanism of bacterial phytase production was carried out by adopting the [55, 65] following method.

2.3.1. The Decision of the Growth Curve of Bacteria Selected

Growth standard curves were made by measuring the Optical Density (OD) value of the selected isolate cultures on the production media at each period. A total of three loops from each pure culture of the selected bacterial isolates were inoculated in 50 ml of media. The suspended bacterial isolates were incubated in a shaker incubator at room temperature with 200 rpm agitation. The OD values were measured every 2 hours with a spectrophotometer at a wavelength (λ) of 600 nm to obtain a series OD using the turbidimetric method. The growth curve was the relationship between the OD value and the incubation time.

2.3.2. Production of Crude Phytases

Phytase production began with making a starter by inoculating three loops of pure culture isolate, from the medium to slant into 50 mL PPM medium, then incubated at room temperature with 100 rpm agitation until the bacteria reached the logarithmic phase, and stored as a starter during phytase production. Then 5 mL of the starter culture suspension was inoculated into 1000 mL of new PPM medium, and divided into four portions of 250 mL each, then incubated at 37°C for 1 x 24 hours with a shaker incubator at 100 rpm. The culture of bacterial cells on the production medium, which has been incubated produced metabolites, and was centrifuged at 5000 rpm for 35 mins at 4°C. The supernatant was obtained, and a crude phytase was separated from the precipitates, then prepared for its activity measurement.

2.3.3. Determine Crude Phytase Activity

2.3.3.1. Preparation of Molybdate-Vanadate Reagent

Preparation of molybdate-vanadate reagentwas carried out by mixing ammonium heptamolybdate solution (20 g/400 mL) and ammonium monovanadate solution (1 g/300 mL) into 140 mL concentrated HNO3, then diluting to one litre.

2.3.3.2. Preparation of Standard Phosphate Solution

Preparation of standard phosphate solution was carried out by dissolving 0.3834 g KH2PO4 in 100 ml of distilled water, then diluted 100 times, fot each millilitre of a solution containing 0.03834 mg KH2PO4.

The standard series were created by taking 0, 0.25, 0.50, 0.75, 1.00, 2.00, 3.00, and 4.00 mL of the solution. Then each was added with 6.25 mL of molybdate vanadate, left to stand for 10 mins, diluted with distilled water to 25 mL, and a series of standard phosphate solutions were obtained.

2.3.3.3. Measurement of Phytase Activity

First, 0.15 mL of crude phytase was incubated with a substrate containing 2 mL of Na-Phytate, 2 mL of $CaCl_2$, and 0.6 mL of 0.1 M Tris-HCl buffer solution pH 7, at room temperature for ± 30 mins. After this, the reaction was stopped by adding 0.75 mL of 5% TCA, and 1.5 mL of the molybdate-vanadate reagent. Then the absorbance was measured using a UV-Vis spectrophotometer at a wavelength of 700 nm. The absorbance value obtained was analyzed with the amount of phosphate content (PO_4^{3-}) formed (Unit/ml) in the crude extract solution of the enzyme using the linear regression equation, from the standard phosphate curve. One specific unit of enzyme phytase (FTU) is defined as the amount of enzyme that catalyzes the formation of 1 μmol of inorganic phosphate per minute under test conditions.

2.3.4. Optimation of Temperature and pH of Maize Plant Endophytic Phytase Activity

The optimization of the activity of each phytase produced by the four endophytic bacteria was carried out by measuring their activity at exposure to temperature combined with pH treatment. The extracellular phytase was incubated in Na-acetate buffer with temperature variations of 20, 30, 40, 50, and 60oC at various pH (2, 3, 4, 5, 6 and 7) for 10 mins. The optimum activity was indicated by the number of enzymes that catalyze the formation of 1 μmol of inorganic phosphate at the combination of pH and temperature tested.

2.4. In Vitro Hydrolysis of Phytate in Feed by Phytase from Endophytic Bacteria of the Maize Plant

This stage aims to determine the effectiveness and optimum dose of endophytic bacteria phytase from plant Maize in hydrolyzing feed phytate. This stage used phytases that have been produced and optimized in the previous step. The phytate in feeds without and with the phytase administration at various levels were measured using modified methods (Buddrick et al. 2014) and (Ishiguro et al. 2003).

2.4.1. Provision of Phytase in Feed

The feed of 0.5 g sterile was placed in Erlenmeyer then added with 50 mL of distilled water. Several phytases (an appropriate level of treatment) were incubated with an incubator shaker at 100 rpm for 3 hours at 40°C, and filtered with filter paper, then the analysis of phytate levels was carried out. The experimental design used was a completely randomized model with six treatments and four replications with the following details: T0 = feed + distilled water; T1 = feed + distilled water + 500 FTU phytase, T2 = feed + distilled water + 750 FTU phytase, T3 = feed + distilled water + 1000 FTU phytase, T4 = feed + distilled water + 1250 FTU phytase, and T5 = feed + distilled water + 1500 FTU phytase.

2.4.2. Preparation of Ca-Phytate Standard Curve

Five test tubes each inserted with 0.0, 0.1, 0.2, 0.3, 0.4, and 0.5 mL of 1.1 mM Ca-phytate solution. Then distilled water was added for all the tubes to have a volume of 0.5 mL. Furthermore, the solution was prepared for absorbance measurement at $\lambda = 465$ nm, using 1 mL of $12H_2O$ $FeNH_4(SO4)$ and 0.9 mL of 0.5 M HNO_3 in each tube, covered with aluminium foil, and immersed in boiling water for 20 mins. After cooling to room temperature, the solution was added with 5 mL of $C_5H_{11}OH$ and 0.1 mL of NH_4SCN. Then homogenized by shaking the tube slowly. Exactly 15 mins after the addition of the NH_4SCN solution into the test tube, it was measured for absorbance by a spectrophotometer at $\lambda = 465$ nm. And Amyl alcohol was used as a blank solution. The data were then used to create a

Ca-phytate standard curve showing the relationship between the amount and the absorbance of phytate. The equation obtained was used to calculate the amount of phytate in the solution. The linear regression equation used was as follows: Y = a + bx (Y = absorbance of the phytate solution, x = the number of phytates in each phytate solution).

2.4.3. Extraction and Measurement of Phytate Levels

The feed filtrate suspension + phytase was added to 50 mL of 0.5 M HNO_3, and incubated for 3 hours on a shaker incubator at room temperature, then filtered and the phytate content in the obtained filtrate was analyzed. A total of 0.05 mL of filtrate was inserted into the test tube, and preparation was made for absorbance measurements at λ = 465 nm, as in the Ca-phytate standard curve. The phytate content in the dry test material was calculated by substituting the absorbance value obtained by the equation of the Ca-phytate standard curve.

3. RESULTS AND DISCUSSION

3.1. Isolation, Screening, and Characterization of Phytase-Producing Endophytic Bacteria from Maize Plants

Isolation of endophytic bacteria from *Zea mays* L. was carried out using liquid and solid Luria Bertani media. Cultivation using the liquid culture method as an initiation medium was carried out using a dilution technique. The growing liquid culture was then inoculated on the agar plate LB medium with the pour plate method to grow bacteria from the culture as colonies. All the colonies of growing isolates were differentiated based on their appearance. However, it was not yet sure whether the isolates obtained were of the same or different species. The colonies of different isolates were then purified on the same media to obtain the types of bacteria that did not mix with other species/strains.

Purification was conducted by sampling a separate loop of bacterial colonies, and scratched on a similar solid medium which was then incubated under the same conditions. From the growing colonies, re-scratching was carried out on the new solid medium up to eleven times, in order to obtain genuinely pure isolates. A total of 28 isolates of endophytic bacteria were obtained from the roots, stems, leaves, and seeds of the Maize plant. The obtained isolates were screened using cultivation and selective media for phytase-producing bacteria, namely PSM by being spotted on agar plates simultaneously. The addition of 0.4% Na-phytate to the PSM, which was initially clear and yellowish, caused the media to become cloudy and milky white. The bacteria that produces good phytase showed a clear zone around the isolated colony, and this is an indication of the enzymatic reaction of phytate hydrolysis contained in the media. Therefore, the wider the clear zone formed the higher the phytase quantity.

Figure 1. Phytatic index (PI) of endophytic bacterial isolates from the maize plant: (a) Isolate from roots (HF.7) with PI value of 1.38, (b) Isolate from the stem (HF.8) with PI value of 1.31, (c) Isolate from seeds (HF.18) with PI value of 1.36, (b) Isolate from the leaf (HF.28) with PI value of 1.23.

All cultivated isolates showed growth on selective medium, and 11 of them indicated the ability to produce good phytase with a Phytatic index (PI) of more than a value of one. The Isolates that grow and form clear zones on the screening medium are phytase-producing bacteria. These bacteria have a phytase gene that was successfully expressed in culture through phytase induction, which depends on two conditions, namely the availability of phytate and the absence of phosphate in the media [57].

Table 1. Characteristics of the colony, cell, Gram, and biochemical activity of phytase-producing endophytic bacterial isolates from the maize plant

Morphological characteristics of isolate colonies				
	HF.7	HF.8	HF.16	HF.28
Colour	yellowish-white	white	white	yellowish-white
Size	moderate	moderate	moderate	moderate
Shape	*irregular*	*circular*	*irregular*	*irregular*
Elevation	*raised*	*raised*	*raised*	*raised*
Edge	*serrate*	*serrate*	*undulate*	*undulate*
Surface	rough	rough	rough	rough
Cell shape and Gram nature of isolate				
Cell shape	basil	basil	coccus	basil
Gram nature	negative	negative	negative	negative
Biochemical assay results of isolates				
TSIA	-	+	+	+
H2S	-			-
Motility	-	+	+	-
Catalase	+	+	+	+
Indol	-	-	-	-
Methyl-Red	-	+	-	-
Voger-Proskauer	-	+	+	+
Citrate	+	+	+	+
Urease	-	-	+	+
Lactose	-	+	+	+
Mannitol	+	+	+	+
Glucose	+	+	+	+

The characteristics of the selected isolates, as shown in Table 2, were known through macroscopic and microscopic observations. The macroscopic observations include remarks of size, pigmentation, shape, elevation, surface, and colony margins. The microscopic observations include cell shape and Gram's nature. The results of Gram staining of the four selected phytase-producing endophytic bacteria showed Gram-negative characteristics.

The differences in bacterial Gram nature occurred as a result of variation in binding capacity and the dyeing process. This was caused by differences in the structure of the cell wall between Gram-positive and negative bacteria. Gram-positive have a thicker peptidoglycan layer than the negative bacteria, therefore, the colour of Crystal violet adheres firmly to it [58].

3.2. Identification of Phytase-Producing Maize Plant Endophytic Bacteria Using a Molecular Approach

Information on the morphological characteristics of the four isolates were different from each other, and became the basis for the identification of a different species. The identification of selected endophytic bacteria was carried out using a molecular approach through analysis of the 16s rRNA gene. The 16s rRNA gene is present in the 30s ribosome subunit and found in all prokaryotes. It has a relatively large number of nucleotides, while some bases are sustainable and arranged as a universal primer to amplify an organism's 16s rRNA gene. The four bacterial isolates were determined using universal primers, namely forward 63F and reverse 1387R, for the 16S-rRNA sequence.

The 16s rRNA gene analysis for the identification of microorganisms using a molecular approach consists of four main processes, namely the extraction/isolation of DNA, PCR, electrophoresis, and sequencing. DNA extraction is the process of separating it from other cell components in order to obtain a pure isolate. The next stage was the amplification process using PCR, which aims to multiply a DNA band in vitro. In the process, there was a chain reaction, namely denaturation, annealing, and elongation. The PCR process was carried out in 35 cycles for ± 2 hours. The forward primers initiated the synthesis of DNA strands from the 5 '-------- 3' end, while the Reverse initiated the synthesis of DNA strands from the 3 '-------- 5' end. The function of the template DNA in the PCR process was a template for the formation of the same new DNA molecule. The chromosomal DNA profiles that have been isolated and multiplied by PCR were analyzed using 1% agarose electrophoresis. The electropherogram from the electrophoresis of the chromosomal DNA of the four isolates (Figure 2), showed the presence of a single thick band produced by each chromosomal DNA of the isolates. From the electrophoresis, results were obtained in the extended base pair (bp), each sample with the help of markers.

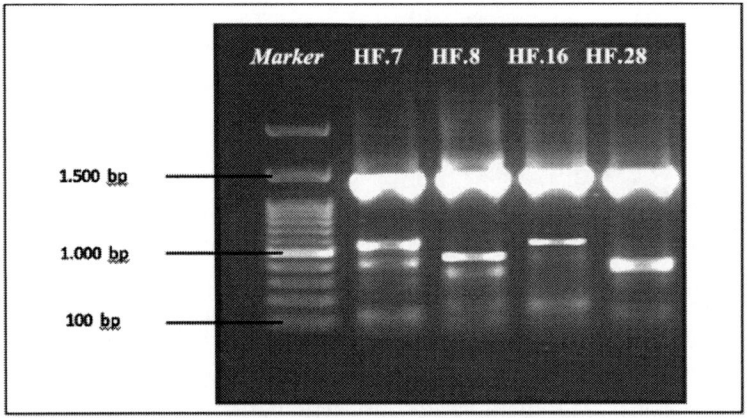

Figure 2. The electrophoregram of the 16S-rRNA gene amplification product of the selected isolates: HF.7 = ± 1000 bp, HF.8 = ± 900 bp, HF.16 = ± 1000 bp, and HF.28 = ± 900 bp.

The basic principle of electrophoresis technique is the separation of charged components or DNA molecules in an electric field. The DNA molecule is separated based on the rate of migration by the electromotive force in the gel matrix. The DNA molecule sample is placed in a well on a gel that is placed in a buffer solution (TBE), then an electric current flow.

The DNA molecule move in the gel matrix towards one of the electric poles according to the charge of the DNA molecule. The direction of the DNA molecules movement toward the positive electrode, was due to the negative charge on the framework of its sugar-phosphate. In order to keep the rate of movement of the DNA molecule strictly based on size, the substance sodium hydroxide is used to keep the DNA straight [54].

After the electrophoresis process, staining was carried out for the separated sample molecules to be clearly observed using ethidium bromide. And the sample molecules glowed in ultraviolet light. The bands in the different stripes of the gel appeared after the dyeing process, representing each lane in the movement direction of the sample from the gel "well". The bands that were equidistant from the gel well at the end of the electrophoresis, contain molecules that moved in the gel during the process at the same speed, meaning that the molecules have the same size. Markers which are molecular mixtures of different sizes were use to determine the size of the molecules in the sample band by electrophoresis, and the markers on the strips in the gel were parallel to the sample.

The bands in the visible marking strip were compared with the sample bands to determine their size. The distance of the band from the gel well was inversely proportional to the logarithm of the molecular size. Based on the results of electrophoresis, DNA isolates of HF.8 and HF.28 showed bands that were perforated and parallel to the markers around ± 900 bp. This indicated that the amplified gene fragment was ± 900 bp in size. The DNA isolates of HF.7 and HF.16 showed separated and parallel bands with markers of around ± 1,000 bp, indicating that the amplified gene fragments were ± 1,000 bp in size.

The 16S-rRNA gene of PCR products was sequenced in the 1st BASE INT Malaysia sequencing. The cluster analysis on sequences was carried out with the online BLAST program from NCBI. The results of the PCR product sequencing of each isolate were intact DNA nucleotide base sequences. Based on the BLAST analysis, the results of the homology of the four selected isolates were as shown in Tables 2, 3, 4, and 5.

Table 2. Homology of nucleotide base sequences of HF.7 isolates

Description	Total score	% of Identities
Burkholderia lata strain 383	1493	99%
Burkholderia contaminans strain J2956	1493	99%
Burkholderia cepacia strain 717	1493	99%
Burkholderia latens strain R-5630	1487	99%
Burkholderia territorii strain LMG 28158	1485	99%
Burkholderia metallica strain R-16017	1482	99%
Burkholderia arboris strain R-24201	1482	99%
Burkholderia cepacia strain ATCC 25416	1482	99%
Burkholderia vietnamiensis strain TVV75	1482	99%
Burkholderia cenocepacia strain LMG 16656	1480	99%

Table 3. Homology of nucleotide base sequences of HF.8 isolates

Description	Total score	% of Identities
Enterobacter cloacae subsp. strain ATCC 2373	1593	99%
Enterobacter cloacae strain DSM 30054	1580	99%
Enterobacter cloacae strain NBRC 13535	1580	99%
Enterobacter cloacae strain 279-56	1580	99%
Enterobacter cloacae subsp. strain LMG 2683	1576	99%
Enterobacter ludwigii strain EN-119	1572	99%
Pantoea agglomerans strain JCM1236	1572	99%
Enterobacter cloacae subsp. strain LMG 2683	1570	99%
Enterobacter cloacae strain ATCC 13047	1559	98%
Enterobacter kobei strain JCM 8580	1557	98%

Table 4. Homology of nucleotide base sequences of HF.16 isolates

Description	Total score	% of Identities
Enterobacter ludwigii strain EN-119	1455	98%
Enterobacter cloacae subsp. strain ATCC	1435	98%
Enterobacter cloacae subsp. strain LMG	1430	98%
Enterobacter kobei strain CIP 105566	1428	98%
Enterobacter cloacae strain DSM 30054	1424	98%
Enterobacter cloacae strain NBRC 13535	1421	98%
Enterobacter kobei strain JCM 8580	1421	98%
Enterobacter cloacae strain 279-56	1421	98%
Leclercia adecarboxylata strain NBRC	1421	98%
Pantoea agglomerans strain JCM1236	1419	98%

Table 5. Homology of nucleotide base sequences of HF.28 isolates

Description	Total score	% of Identities
Pantoea stewartii subsp. indologenes strain CIP	2069	97%
Pantoea stewartii strain LMG 2715	2050	97%
Pantoea stewartii strain ATCC 8199	2050	97%
Pantoea stewartii strain LMG	2023	96%
Pantoea allii strain BD 390	1971	95%
Pantoea ananatis strain 1846	1960	95%
Raoultella electrica strain 1GB	1954	95%
Pantoea ananatis strain LMG 2665	1949	95%
Pantoea anthophila strain LMG 2558	1943	95%
Raoultella ornithinolytica strain ATCC 31898	1943	95%

The results of microorganism homology obtained from the BLAST analysis provided the highest similarity information to the nucleotide sequences of the isolates, these results were confirmed by the DNA sequences of microorganisms from around the world that were deposited in the NCBI GenBank database. The important information from the BLAST results was in the form of a total score and the percentage of identities. The total score was the sum of the alignments of all segments of the database sequence that matched the nucleotide sequence. This value indicated the accuracy of the sequence value in the form of an unknown nucleotide with those contained in the GenBank. The higher the score obtained, the higher the homology level of the two sequences. The Identities obtained had the highest value between the query and the aligned database sequence [59, 60].

The analysis of the DNA sequences' similarities of the four selected isolates with those contained in the GenBank showed that, endophytic bacterial isolates from Maize root with code HF.7 have up to 99% similarity with ten species from the genus Burkholderia listed in the NCBI database. Based on the highest score, three isolates showed the same value, namely 1493. This indicated that the endophytic bacteria with code HF.7 have genetic similarities with the species of *Burkholderia lata, B. contaminans,* and *B. cepacia*. Specifically, it was uncertain that HF.7 isolate is either a specie of the three genera, therefore, endophytic bacterial isolate, which was obtained from Maize root with code HF.7, was designated as *Burkholderia* sp. strain HF.7. The endophytic bacterial isolate from the stem, coded HF.8 has 99% similarity with *Enterobacter cloacae subsp. strain ATCC 2373*. Those isolated from leaves coded HF.16 has 98% similarity with *Enterobacter ludwigii strain EN-119,* and endophytic bacterial isolates from seeds with code HF.28, has 97% similarity with *Pantoea stewartii subsp. indologenes strain CIP.*

Meanwhile, the exploration of the ability of the genus Burkholderia, Enterobacter cloacae, E. ludwigii, and Pantoea stewartii in producing phytase is still minimal. However, several reported bacterial species are closely related to these isolates, for example, Burkholderia sp. strain a13 and Pantoea agglomerans. The ability of bacteria to produce phytase is primarily determined by phytase induction which depends on two conditions, namely the availability of the substrate (Na-Phytat or Ca-phytate) and the absence of inorganic phosphate in the media because phytase is an inductive enzyme [57].

3.2.1. *Burkholderia sp.*

Burkholderia genus is widespread in various ecological environment, however, it is commonly found in soil and shows the interaction of non-pathogenic to crops. It is also able to dissolve minerals in the soil by producing organic acids, as well as increasing the availability of nutrients for plants, making it very promising to be used in the field of biotechnology. Burkholderia is a genus of endophytic bacteria that is often found in rice, maize, and sugarcane and is capable of producing bioactive compounds, one of which is used as an antimicrobial compound [61-63]. As a general characteristic of Burkholderia, the endophytic bacterial isolate HF.7 is a Gram-negative, rod-shaped, non-motile, and aerobic. Their bacteria colony is moist and has yellow pigment, and grow well at of 30°C-37°C. The taxonomy of the Burkholderia is as follows:

Domain: Bacteria
Phylum: Proteobacteria
Class: Beta Proteobacteria

Order: Burkholderiales
Family: Burkholderiaceae
Genus: *Burkholderia* sp.

Other biochemical properties of HF.7 isolates are the ability to assimilate mannitol, glucose, and citrate, without urease and tryptone activity. This resembles the physiological characteristics of one of the Burkholderia species, namely *B. lata* [64]. These bacteria have catalase and lysine decarboxylase activity, without tryptophanase, arginine, dihydrolase, or urease activity in their metabolism. Several strains of Burkholderia lata assimilate D-glucose, D-mannose, D-mannitol, N-acetylglucosamine, D-Burkholderiagluconate, L-malate, and citrate. At the same time, the assimilation of maltose, L-arabinose, and phenylacetate depends on the strain of *Burkholderia lata* [64].

From previous reports, the genus Burkholderia bacteria is capable of associating with rhizosphere plants and contribute to plant growth, by freeing phosphate from the soil organic compounds, such as phytate [61]. Although bacterial strains of the genus Burkholderia have not been widely reported to integrate phytate, there have been reports of the characteristics of phytase produced by *Burkholderia* sp. strain a13. Phytase produced by this strain showed a specific activity of 4.1 U/mg. The optimum conditions of the temperature and pH for *Burkholderia* sp. strain a13 in producing phytase are 45-55°C, pH 4.5 and the product is stable up to 4°C [61, 65, 66].

3.2.2. Enterobacter Cloacae

Enterobacter cloacae belongs to the Enterobacteriaceae group with Gram-negative characteristics in the form of bacilli. According to [67], the bacteria *Enterobacter* sp. produces phytase. However, there have been no report on the species of *Enterobacter cloacae* producing phytase. The taxonomy of the Enterobacter cloacae bacteria is as follows:

Domain: Bacteria
Phylum: Proteobacteria

Class: Gammaproteobacteria
Order: Enterobacteriales
Family: Enterobacteriaceae
Genus: Enterobacter
Species: Enterobacter cloacae

Several studies reported that the endophytic bacteria *Enterobacter cloacae* were shown to increase nitrogen fixation in rice plants. In addition to producing IAA hormone and having the ability to increase nitrogen fixation, they also produce the enzyme L-Histidine Decarboxylase (HDC) [67, 68].

3.2.3. Enterobacter Ludwigii

Enterobacter ludwigii is an endophytic bacterium that is included in the general characteristics of the genus enterobacter. It is a Gram-negative bacterium, bacillus, motile, and able to ferment. The taxonomy of the Enterobacter ludwigii bacteria is as follows:

Domain: Bacteria
Phylum: Proteobacteria
Class: Gammaproteobacteria
Order: Enterobacteriales
Family: Enterobacteriaceae
Genus: Enterobacter
Species: Enterobacter ludwigii

It was previously shown that the bacteria *Enterobacter* sp. produce phytase. However, studies reporting that *Enterobacter ludwigii* species produce phytase have not been found. Only a few reported on the potential of the bacteria *Enterobacter ludwigii* [69-71]. Several studies have reported the abilities possessed by *Enterobacter ludwigii*, including having activity as Plant Growth Promoting Bacteria (PGPB). Those isolated from the rhizosphere of *Lolium perenne* L. grass showed phosphate solvent activity, nitrogen-fixing, and producing growth hormone IAA [72, 73].

3.2.4. Pantoea Stewartii

Pantoea stewartii is a gram-negative bacterium and non-motile. There is a genus of Pantoea which is pathogenic to plants, while some are beneficial (in association). Species from Pantoea stewartii sp indologenes are associated with sorghum grass. The classification of the Pantoea stewartii bacteria is as follows:

Domain:	Bacteria
Phylum:	Proteobacteria
Class:	Gammaproteobacteria
Order:	Enterobacteriales
Family:	Enterobacteriaceae
Genus:	Enterobacter
Species:	*Pantoea stewartii*

Several species of the genus Pantoea are reported to produce phytase, including *Pantoea agglomerans* [74], which has been shown to reduce phytate content in feed using the phytase it produced [75, 76]. [77]. Also, previous research reported that this specie was successfully isolated from the soil of the Republic of Tatarstan, Russia, based on the high activity of phytate decomposers which stores 99% 16S rRNA nucleotide sequence similar to the *Pantoea* sp. Moreover, [53] the ability of *Pantoea stewartii* ASUIA271 to produce phytase was explored, which was triggered by the high organic phytate content in rice husks at various experimental temperatures.

3.3. Production and optimIzation of Phytase Activity of Endophytic Bacteria from Maize Plants

3.3.1. Phytase Crude Extract Production

Phytase production began with the determination of the growth curves of the four selected endophytic bacterial isolates. It was crucial to observe the growth/survival pattern of bacteria for the optimum phase of the highest phytase activity to be found in each of these isolates. The growth curves of the four selected endophytic bacterial isolates were shown in Figure 8. The growth curve is the relationship between the Optical Density (OD) value and the incubation time. OD values were measured every 2 hours with a spectrophotometer at a wavelength (λ) of 600 nm, to obtain a series OD using the turbidimetric method.

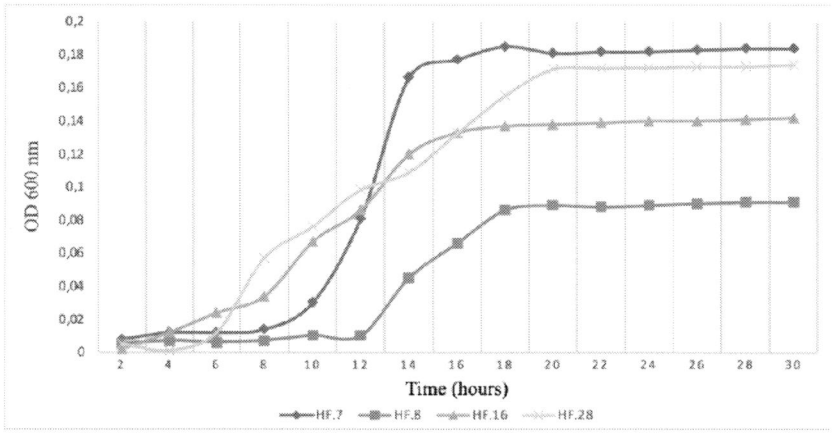

Figure 3. Growth curve of endophytic bacterial isolates from *Zea mays* L. (a) *Burkholderia* sp. strain HF.7, (b) *Pantoea stewartii* strain HF.8, (c) *Enterobacter ludwigii* strain HF.16, (d) *Enterobacter cloacae* strain HF. 28.

In general, enzymes are produced during bacterial growth. They reach their highest activity at the end of the exponential or log phase. in the previous stage, namely the adaptation (lag phase), the individual bacteria growing into adults do not undergo cell division, however, adapt to the growth environment. The growth curve (Figure 3) showed that the log phases of the four bacterial isolates were achieved at 8, 12, 4, and 4 hours for isolates HF.7, HF.8, HF.16, and HF.28, respectively. The incubation time was the basis required as a starter, in order for the phytase production to last until the end of the log phase, and also extendable to a half of the stationary phase. At the end of the log phase, there was a peak increase in the number of cells because each active cell multiplies, while in the stationary stage there was no multiplication of bacterial cells [78].

Based on the growth curves of the four selected isolates, phytase production was carried out by inoculating each of the four isolates starter culture as much as 5 mL into 250 mL sterile PPM media, and incubated at 28°C for 18, 18, 16, and 20 hours for isolates HF.7, HF.8, HF.16, and HF.28, respectively, using an incubator shaker at a speed of 100 rpm. The culture was then centrifuged at 5000 rpm for 35 mins at 4°C. The supernatant obtained by separating the precipitate was crude phytase, then its activity was measured.

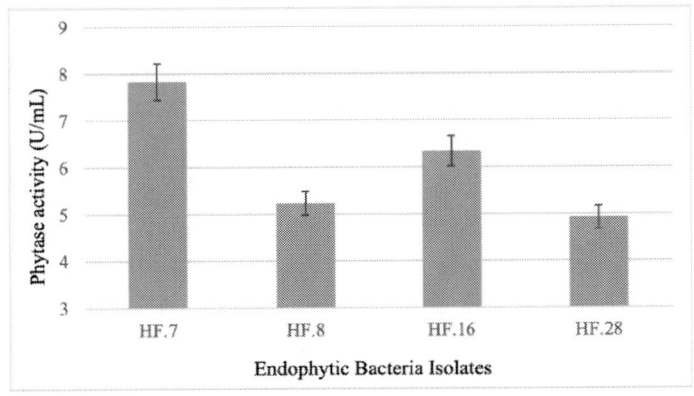

Figure 4. Phytase activity of the endophytic bacterial isolate *Zea mays* L. (a) *Burkholderia* sp. strain HF.7, (b) *Pantoea stewartii* strain HF.8, (c) *Enterobacter ludwigii* Strain HF.16, (d) *Enterobacter cloacae* strain HF. 28.

The activity of phytase was known by measuring the absorbance value of the crude phytase at a wavelength of 700 nm. Then the amount of phosphate content formed (FTU/mL) in the enzyme crude extract solution was analyzed by substituting the absorbance value using a linear regression equation from the phosphate standard curve [49]. One phytase unit (FTU) is defined as the amount of enzyme that catalyzes the formation of 1 µmol of phosphate per minute. The activity of the crude phytase extract produced by the four selected endophytic bacteria was presented in Figure 4.

Based on the analysis of the variance, the phytase produced by endophytic bacterial from *Zea mays* L. with code HF.7 showed the highest activity. Which was significantly different ($P < 0.01$) compared to the phytase produced by the three other endophytic bacteria isolates. An interesting observation in the endophytic bacteria was that, the best phytase producer was obtained from their Maize root. This was due to the phytate content in its root area, which triggers the emergence of phytase-producing endophytic bacteria, as a result of inositol phosphate, which is widely distributed in Maize roots.

The phytase activity produced by the four selected endophytic bacterial, between 4.9-7.8 FTU/mL, showed higher value compared to those generated by *Burkholderia* sp. strain a13 (4.1 FTU/mL), *Bacillus cereus* ASUIA 260 (1.160 FTU/mL), and Bacillus subtilis AP-17 (0.0296 FTU/mL) [57]. Similarly, the phytase produced by three *Bacillus cereus* strains isolated from the volcanic ash of Mount Merapi includes 0.1071 FTU/mL, 0.1020 U/mL, and 0.0874 FTU/mL [79] [53]. It was also reported that the phytase activity of *Staphylococcus lentus* ASUIA 279 was 1.913 FTU/mL. [80]. Moreover, the phytase activity of the three strains of *Bacillus cereus,* isolated from water and mud samples of the Sikidang Dieng crater were 0.32893 FTU/mL, 0.324953 FTU/mL, and 0.32182 FTU/mL.

3.3.2. The Optimization of Temperature and pH of Phytase Activity from Endophytic Bacteria of Maize Plants

This test was conducted to determine the optimum phytase activity produced from the four selected maize plant endophytic bacteria at the same temperature and pH as well as the protease activity of the poultry digestive tract in vitro. The determination of the optimum temperature and pH was measured by observing the activity of the phytase crude extract, exposed to varying temperatures (30, 40, 50, 60, and 70°C) for 10 mins. The pH optimation test was carried out by incubating the crude extract of extracellular phytase in Na-acetate buffer with varying pH (2, 3, 4, 5, 6, and 7).

The results of the extracellular phytase optimization with variations in temperature and pH produced by each isolate were shown in Figure 5. The phytase of *Burkholderia* sp. strain HF.7 showed activity in all variations in pH treatment. Extracellular phytase activity of *Burkholderia* sp. strain

HF.7 as observed from the treatment of pH 2, showed an increase in activity from pH 3 to 4, and at pH 5 to pH 6, and decreased activity at pH 7 [81]. The changes in activity at different pHs were caused by the occurrence of intramolecular changes of enzymes caused by ionization to bind and release protons (hydrogen ions) in amino, carboxyl, and other functional groups. When the difference was too large, it results in the denaturation of the enzyme, and its activity was lost [82]. The results of this measurement were in line with the nature of phytase, which are heterologous group of enzymes, hydrolyse phosphate esters, and optimal at low pH. At neutral and alkaline pH, catalytic activity decreases. This is due to the structural instability of the enzyme protein molecules, which causes structural changes in these pH conditions [83, 84].

The pH condition of the digestive tract of poultry, especially chicken, is 4.5 in the crop, 4.4 in the proventriculus, 2.6 in the gizzard, 5.7-6.0 in the duodenum, 5.8 in the jejunum, 6.3 in the ileum, 6.3 in the colon, and 5.7 in the caeca. The extracellular phytase of the four endophytic bacteria of the Maize plant showed stable activity at pH 4-6 [85]. Therefore, the phytase produced by the endophytic bacteria of the maize plant is active in the digestive tract of the poultry.

The determination of extracellular phytase activity of maize plant endophytic bacteria against a combination of temperature and pH variations as in Figure 5, showed that increasing temperature causes an increase in activity until it reaches the optimum point of 40°C. The rise in temperature decreased the phytase activity, as observed at 50°C and continues to decline until a temperature of 70°C. At first, with increasing temperature, the enzyme reaction speed increases due to the rise in kinetic energy, which accelerated the vibrational, translational, and rotational motion of the enzyme and the substrate, increasing their chance to react. On exposure to temperatures higher than the optimum, the protein and the substrate underwent a conformational change. This caused the reactive group to unmatch the active side, or experience obstacles in entering the active site of the enzyme, in order to significantly affects its catalytic activity [83].

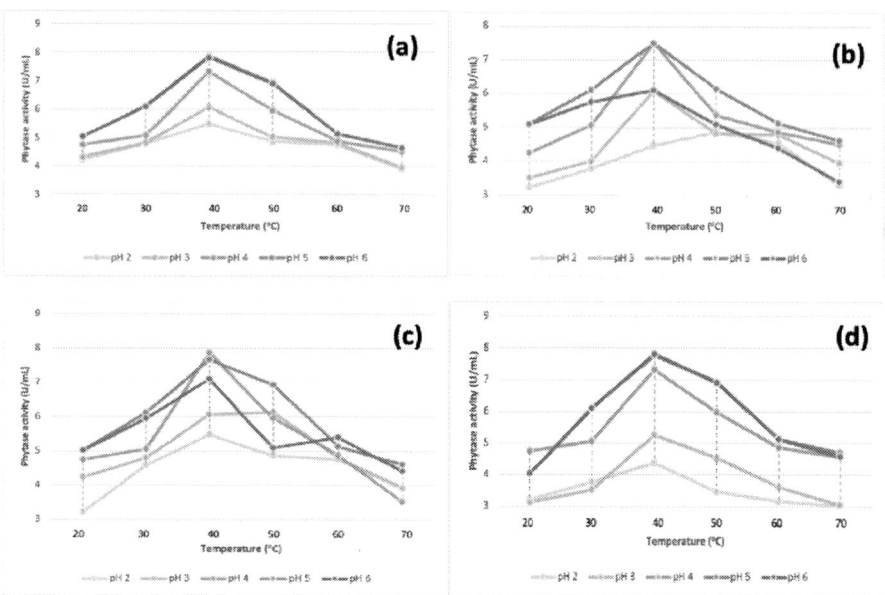

Figure 5. Phytase activity of maize plant endophytic bacteria at a combination of temperature and pH variations: (a) *Burkholderia* sp. strain HF.7, (b) *Pantoea stewartii* strain HF.8, (c) *Enterobacter ludwigii* Strain HF.16, (d) *Enterobacter cloacae* strain HF. 28.

The optimum temperature and pH of the enzyme depends on its type and source. Based on the activity measurement, it was known that the optimum temperature and pH of phytase produced by *Burkholderia* sp. strain HF.7 and Enterobacter cloacae strain HF.28 was 40°C and pH 6. Pantoea stewartii strain HF.8 generated 40°C and pH 5. Enterobacter ludwigii Strain HF.16 produced 40°C and pH 4. Temperature variations and the optimum pH of phytase activity produced by other bacteria have also been widely reported [86], namely those obtained from *B. subtillis* (natto) N-77 were 60°C and pH 6.0-6.5, *Enterobacter* sp. optimum were pH 7.5 and 50°C. Phytase is also produced by *Aspergillus niger* (58°C, pH 5.5), *Schwanniomiyces castellii* (7.7°C, pH 4.4), and *Klebsiella aerogenes* (45°C, pH 7.0).

3.4. In Vitro Hydrolysis of Phytate in Feed by Phytase from Endophytic Bacteria of the Maize Plant

This stage aims to determine the level of feed phytate hydrolysis by Maize plant endophytic bacteria at various levels, in order to obtain the best phase for in vivo supplementation. This stage used phytase, which was produced and optimized in the previous step. The parameters measured were phytate levels in feed without and with supplementation from the endophytic bacteria. The measurement of phytate levels was carried out on sterile feed samples, added with 50 mL of distilled water and several phytases (according to treatment level) for 3 hours at room temperature and filtered. The six phytase levels were used to measure the hydrolysis in feed by in vitro bacteria, namely 0, 500, 750, 1000, 1250, and 1500 FTU [87, 88]. The effect of adding phytase at various levels in basal feed was presented in Figure 6.

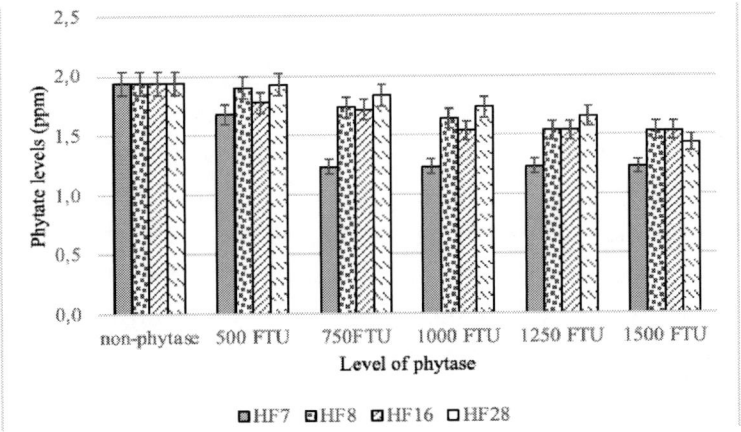

Figure 6. The phytate levels of the endophytic bacteria isolates hydrolysis of maize plants at different stages: T0 (non-phytase), T1 (500 FTU phytase), T2 (750 FTU phytase), T3 1000 FTU phytase, T4 (1250 FTU phytase), dan T5 (1500 FTU phytase).

Based on the determination of phytate levels as shown in Figure 6, the addition of the treatment showed a decrease in feed up to 60%, as shown in Figure 7. The results of variance analysis showed that the addition of phytase had a significant effect in reducing the feed level. This indicated that the phytase of *Burkholderia* sp. strain HF.7 added to feed with levels of 500, 750, 1000, 1250, and 1500 FTU/kg reduced their phytate.

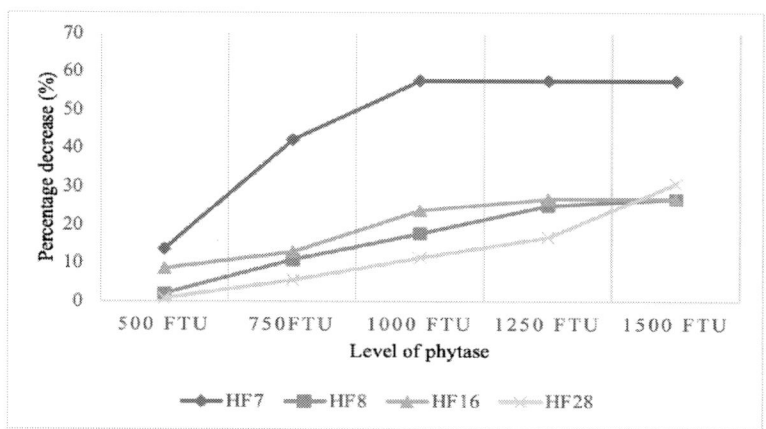

Figure 7. Percentage reduction by in vitro hydrolysis in feed by endophytic bacterial phytase of the maize plant.

The addition of 500 FTU phytase was able to reduce 26.53% of phytate levels compared to without it. Meanwhile, the addition of 750 FTU phytase significantly increased the decrease in phytate levels of feed by 70.80%. This data showed that the optimum dose of phytase produced by *Burkholderia* sp. strain HF.7 is at the level of 750 FTU. The addition of phytase concentrations exceeding 750 FTU showed a decrease in phytate levels, which was not significant. An imbalance between the substrates produced this result. For optimum phytate degradation, the ratio of its amount between the enzyme and the substrate should be balanced. Basically, the greater the enzyme concentration, the faster the reaction takes place, i.e., the enzyme concentration is directly proportional to the reaction speed. However, with a limited number of substrates, the degradation rate is lower. The phytase produced by three other endophytic bacteria showed that it needed higher quantities to hydrolyse phytates in feed, where each was able to reduce by 20% at the 1500 FTU level.

CONCLUSION

This study shows other significant role of *Zea mays* L. besides being a food source. The endophytic bacteria that are symbiotic in the maize plant cycle, are known to play a role in increasing its growth and yield, suppressing contaminant pathogens, dissolving phosphates, or contributing nitrogen. The other potential endophytic bacteria producing a very prospective enzyme, namely phytase, were applied in improving the quality of broiler feed. This research found and identified four types of potential phytase-producing endophytic bacteria from the Maize plant, namely *Burkholderia* sp. strains HF.7, *Enterobacter cloacae* strains HF.8, *Enterobacter ludwigii* strains HF.16, and *Pantoea stewartii* strains HF.28. Sequentially, each was isolated from the roots, stems, leaves, and seeds of maize plants. Also, each of the phytase produced by the four endophytic bacteria hydrolysed phytate which has been tested in vitro at various dose, temperature, and pH. Therefore, this enzyme is proven to have the potential of improving poultry feed quality.

REFERENCES

[1] Tanumihardjo, Sherry, Laura McCulley, Rachel Roh, Santiago Lopez-Ridaura, Natalia Palacios-Rojas, and Nilupa S. Gunaratna. 2020. "Maize Agro-Food Systems to Ensure Food and Nutrition Security in Reference to the Sustainable Development Goals." *Global Food Security* 25: 100-113. Accessed July 12, 2020. doi:org/10.1016/j.gfs.2019.100327.

[2] Ranum, Peter, Juan Pablo Peña-Rosas, and Maria Nieves Garcia-Casal. 2014. "Global Maize Production, Utilization, and Consumption." *Annals of the New York Academy of Sciences* 1312 (1): 105–112. Accessed September 21, 2020. doi:org/10.1111/nyas.12396.

[3] Li, Quanfeng, Jianjun Zang, Dewen Liu, Xiangshu Piao, Changhua

Lai, and Defa Li. 2014. "Predicting Corn Digestible and Metabolizable Energy Content from Its Chemical Composition in Growing Pigs." *Journal of Animal Science and Biotechnology.* 12 (2):186-194. Accessed July 22, 2020. doi:org/10.1186/2049-1891-5-11.

[4] Zhao, F., H. F. Zhang, S. S. Hou, and Z. Y. Zhang. 2008. "Predicting Metabolizable Energy of Normal Corn from Its Chemical Composition in Adult Pekin Ducks." *Poultry Science.* 87 (8): 1603-1608. Accessed July 22, 2020. doi:org/10.3382/ps.2007-00494.

[5] Smith, K. A., C. L. Wyatt, and J. T. Lee. 2019. "Evaluation of Increasing Levels of Phytase in Diets Containing Variable Levels of Amino Acids on Male Broiler Performance and Processing Yields." *Journal of Applied Poultry Research* 28 (2): 253–62. Accessed July 22, 2020. doi:org/10.3382/japr/pfy065.

[6] Singh, Nand Kumar, Dharmendra Kumar Joshi, and Raj Kishor Gupta. 2013. "Isolation of Phytase Producing Bacteria and Optimization of Phytase Production Parameters." *Jundishapur Journal of Microbiology* 6 (5):85-97. Accessed July 22, 2020. doi:org/10.5812/jjm.6419.

[7] Oatway, Lori, Thava Vasanthan, and James H. Helm. 2001. "Phytic Acid." *Food Reviews International* 17 (4): 419–31. Accessed July 22, 2020. doi:org/10.1081/FRI-100108531.

[8] Kies, Arie K., Leon H. De Jonge, Paul A. Kemme, and Age W. Jongbloed. 2006. "Interaction between Protein, Phytate, and Microbial Phytase. In Vitro Studies." *Journal of Agricultural and Food Chemistry* 54 (5): 1753–58. Accessed July 28, 2020. doi:org/10.1021/jf0518554.

[9] Singh, P. K. 2008. "Significance of Phytic Acid and Supplemental Phytase in Chicken Nutrition: A Review." *World's Poultry Science Journal* 64 (4): 553–80. Accessed July 22, 2020. doi:org/10.1017/S0043933908000202.

[10] Dersjant-Li, Y., C. Evans, and A. Kumar. 2018. "Effect of Phytase Dose and Reduction in Dietary Calcium on Performance, Nutrient Digestibility, Bone Ash and Mineralization in Broilers Fed Corn-

Soybean Meal-Based Diets with Reduced Nutrient Density." *Animal Feed Science and Technology* 242 (8): 95–110. Accessed July 29, 2020. doi:org/10.1016/j.anifeedsci.2018.05.013.

[11] Kumar, Vikas, and Amit K. Sinha. 2018. "General Aspects of Phytases." In *Enzymes in Human and Animal Nutrition*, 53–72. Elsevier. Accessed July 29, 2020. doi:org/10.1016/B978-0-12-805419-2.00003-4.

[12] Humer, E., C. Schwarz, and K. Schedle. 2015. "Phytate in Pig and Poultry Nutrition." *Journal of Animal Physiology and Animal Nutrition* 99 (4): 605–25. Accessed July 29, 2020. doi:org/10.1111/jpn.12258.

[13] Bougouin, A., J. A. D. R. N. Appuhamy, E. Kebreab, J. Dijkstra, R. P. Kwakkel, and J. France. 2014. "Effects of Phytase Supplementation on Phosphorus Retention in Broilers and Layers: A Meta-Analysis." *Poultry Science* 93 (8): 1981–92. Accessed July 29, 2020. doi:org/10.3382/ps.2013-03820.

[14] Catalá-Gregori, Pablo, Victoria García, Josefa Madrid, Juan Orengo, and Fuensanta Hernández. 2007. "Response of Broilers to Feeding Low-Calcium and Total Phosphorus Wheat-Soybean Based Diets plus Phytase: Performance, Digestibility, Mineral Retention and Tibiotarsus Mineralization." *Canadian Journal of Animal Science* 87 (4): 563–69. Accessed July 29, 2020. doi:org/10.4141/CJAS07059.

[15] Kornegay, E. T., A. F. Harper, R. D. Jones, and L. J. Boyd. 1997. "Environmental Nutrition: Nutrient Management Strategies to Reduce Nutrient Excretion of Swine." *The Professional Animal Scientist* 13 (3): 99–111. Accessed July 29, 2020. doi:org/10.15232/S1080-7446(15)31861-1.

[16] Nielsen, Per H., and Henrik Wenzel. 2007. "Environmental Assessment of Ronozyme® P5000 CT Phytase as an Alternative to Inorganic Phosphate Supplementation to Pig Feed Used in Intensive Pig Production." *The International Journal of Life Cycle Assessment* 12 (7): 514–20. Accessed July 29, 2020. doi:org/10.1065/lca2006.08.265.2.

[17] Jooste, Michelle, Francois Roets, Guy F. Midgley, Kenneth C.

Oberlander, and Leánne L. Dreyer. 2019. "Nitrogen-Fixing Bacteria and Oxalis-Evidence for a Vertically Inherited Bacterial Symbiosis." *BMC Plant Biology*. Accessed July 29, 2020. doi:org/10.1186/s12870-019-2049-7.

[18] Kandel, Shyam, Pierre Joubert, and Sharon Doty. 2017. "Bacterial Endophyte Colonization and Distribution within Plants." *Microorganisms*. 21(3):116-124. Accessed September 12, 2020. doi:org/10.3390/microorganisms5040077.

[19] Song, Hai-Yan, Aly Farag El Sheikha, and Dian-Ming Hu. 2019. "The Positive Impacts of Microbial Phytase on Its Nutritional Applications." *Trends in Food Science & Technology* 86 (April): 553–62. Accessed July 22, 2020. doi:org/10.1016/j.tifs.2018.12.001.

[20] Lamid, M., A. Al-Arif, O. Asmarani, and S. H. Warsito. 2018. "Characterization of Phytase Enzymes as Feed Additive for Poultry and Feed." In *IOP Conference Series: Earth and Environmental Science*, 137:012009. Accessed September 22, 2020. doi:org/10.1088/1755-1315/137/1/012009.

[21] Farhadi, D., A. Karimi, Gh Sadeghi, J. Rostamzadeh, and M. R. Bedford. 2017. "Effects of a High Dose of Microbial Phytase and Myo-Inositol Supplementation on Growth Performance, Tibia Mineralization, Nutrient Digestibility, Litter Moisture Content, and Foot Problems in Broiler Chickens Fed Phosphorus-Deficient Diets." *Poultry Science* 96 (10): 3664–3675. Accessed July 22, 2020. doi:org/10.3382/ps/pex186.

[22] Miliute, Inga, Odeta Buzaite, Danas Baniulis, and Vidmantas Stanys. 2015. "Bacterial Endophytes in Agricultural Crops and Their Role in Stress Tolerance: A Review." *Zemdirbyste-Agriculture*. 102 (4): 465-478. Accessed September 22, 2020. doi:org/10.13080/z-a.2015.102.060.

[23] Brader, Günter, Stéphane Compant, Birgit Mitter, Friederike Trognitz, and Angela Sessitsch. 2014. "Metabolic Potential of Endophytic Bacteria." *Current Opinion in Biotechnology*. 27 (1): 30-37. Accessed September 22, 2020. doi:org/10.1016/j.copbio.2013.09.012.

[24] Ryan, Robert P., Kieran Germaine, Ashley Franks, David J. Ryan, and

David N. Dowling. 2008. "Bacterial Endophytes: Recent Developments and Applications." *FEMS Microbiology Letters*. 278 (1): 1-9. Accessed September 22, 2020. doi:org/10.1111/j.1574-6968.2007.00918.x.

[25] Obermeier, M. M., and C. A. M. Bogot. 2019. "Prospects for Biotechnological Exploitation of Endophytes Using Functional Metagenomics." In *Endophyte Biotechnology: Potential for Agriculture and Pharmacology*, 164–79. Wallingford: CABI. Accessed September 29, 2020. doi:org/10.1079/9781786399427.0164.

[26] Woyengo, Tofuko A., Adewale Emiola, Augustine Owusu-Asiedu, Wilhelm Guenter, Philip H. Simmins, and Charles M. Nyachoti. 2010. "Performance and Nutrient Utilization Responses in Broilers Fed Phytase Supplemented Mash or Pelleted Corn-Soybean Meal-Based Diets." *The Journal of Poultry Science* 47 (4): 310–15. Accessed September 22, 2020. doi:org/10.2141/jpsa.009124.

[27] Cowieson, A. J., T. Acamovic, and M. R. Bedford. 2006. "Phytic Acid and Phytase: Implications for Protein Utilization by Poultry." *Poultry Science* 85 (5): 878–85. Accessed Juli 24, 2020. doi:org/10.1093/ps/85.5.878.

[28] Zeller, Ellen, Margit Schollenberger, Maren Witzig, Yauheni Shastak, Imke Kühn, Ludwig E. Hoelzle, and Markus Rodehutscord. 2015. "Interactions between Supplemented Mineral Phosphorus and Phytase on Phytate Hydrolysis and Inositol Phosphates in the Small Intestine of Broilers." *Poultry Science* 94 (5): 1018–29. Accessed Juli 12, 2020. doi:org/10.3382/ps/pev087.

[29] Selle, P. H., V. Ravindran, A. Caldwell, and W. L. Bryden. 2000. "Phytate and Phytase: Consequences for Protein Utilisation." *Nutrition Research Reviews* 13 (2): 255–78. Accessed Juli 12, 2020. doi:org/10.1079/095442200108729098.

[30] Angel, R., W. W. Saylor, A. D. Mitchell, W. Powers, and T. J. Applegate. 2006. "Effect of Dietary Phosphorus, Phytase, and 25-Hydroxycholecalciferol on Broiler Chicken Bone Mineralization,

Litter Phosphorus, and Processing Yields." *Poultry Science* 85 (7): 1200–1211. Accessed Juli 12, 2020. doi:org/10.1093/ps/85.7.1200.

[31] Liu, Bing-Lan, Amjad Rafiq, Yew-Min Tzeng, and Abdul Rob. 1998. "The Induction and Characterization of Phytase and Beyond." *Enzyme and Microbial Technology* 22 (5): 415–24. Accessed Juli 12, 2020. doi:org/10.1016/S0141-0229(97)00210-X.

[32] Bhavsar, K., and J. M. Khire. 2014. "Current Research and Future Perspectives of Phytase Bioprocessing." *RSC Adv.* 4 (51): 26677–91. Accessed September 15, 2020. doi:org/10.1039/C4RA03445G.

[33] Lei, Xin Gen, Jeremy D. Weaver, Edward Mullaney, Abul H. Ullah, and Michael J. Azain. 2013. "Phytase, a New Life for an 'Old' Enzyme." *Annual Review of Animal Biosciences* 1 (1): 283–309. Accessed September 15, 2020. doi:org/10.1146/annurev-animal-031412-103717.

[34] Selle, Peter H., and Velmurugu Ravindran. 2007. "Microbial Phytase in Poultry Nutrition." *Animal Feed Science and Technology* 135 (1–2): 1–41. Accessed September 15, 2020. doi:org/10.1016/j.anifeedsci.2006.06.010.

[35] Dessimoni, Gabriel Villela, Nilva Kazue Sakomura, Daniella Carolina Zanardo Donato, Fábio Goldflus, Nayara Tavares Ferreira, and Felipe Santos Dalólio. 2019. "Effect of Supplementation with Escherichia Coli Phytase for Broilers on Performance, Nutrient Digestibility, Minerals in the Tibia and Diet Cost." *Semina: Ciências Agrárias* 40 (2): 767. Accessed September 15, 2020. doi:org/10.5433/1679-0359.2019v40n2p767.

[36] Buchanan, N. P., K. G. S. Lilly, and J. S. Moritz. 2010. "The Effects of Diet Formulation, Manufacturing Technique, and Antibiotic Inclusion on Broiler Performance and Intestinal Morphology." *Journal of Applied Poultry Research* 19 (2): 121–31. Accessed September 15, 2020. doi:org/10.3382/japr.2009-00071.

[37] Lim, Boon Leong, Pok Yeung, Chiwai Cheng, and Jane Emily Hill. 2007. "Distribution and Diversity of Phytate-Mineralizing Bacteria." *The ISME Journal* 1 (4): 321–30. Accessed September 15, 2020. doi:org/10.1038/ismej.2007.40.

[38] Viveros, A., A. Brenes, I. Arija, and C. Centeno. 2002. "Effects of Microbial Phytase Supplementation on Mineral Utilization and Serum Enzyme Activities in Broiler Chicks Fed Different Levels of Phosphorus." *Poultry Science* 81 (8): 1172–83. Accessed September 28, 2020. doi:org/10.1093/ps/81.8.1172.

[39] Vandenberg, G. W., S. L. Scott, P. K. Sarker, V. Dallaire, and J. de la Noüe. 2011. "Encapsulation of Microbial Phytase: Effects on Phosphorus Bioavailability in Rainbow Trout (Oncorhynchus Mykiss)." *Animal Feed Science and Technology* 169 (4): 230–43. Accessed September 28, 2020. doi:org/10.1016/j.anifeedsci.2011.07.001.

[40] Cowieson, A. J., M. R. Bedford, P. H. Selle, and V. Ravindran. 2009. "Phytate and Microbial Phytase: Implications for Endogenous Nitrogen Losses and Nutrient Availability." *World's Poultry Science Journal* 65 (3): 401–18. Accessed September 28, 2020. doi:10.1017/S0043933909000294

[41] Rutherfurd, S. M., T. K. Chung, P. C. Morel, and P. J. Moughan. 2004. "Effect of Microbial Phytase on Ileal Digestibility of Phytate Phosphorus, Total Phosphorus, and Amino Acids in a Low-Phosphorus Diet for Broilers." *Poultry Science* 83 (1): 61–68. Accessed September 28, 2020. doi:org/10.1093/ps/83.1.61.

[42] Raveendran, Sindhu, Binod Parameswaran, Sabeela Beevi Ummalyma, Amith Abraham, Anil Kuruvilla Mathew, Aravind Madhavan, Sharrel Rebello, and Ashok Pandey. 2018. "Applications of Microbial Enzymes in Food Industry." *Food Technology and Biotechnology* 56 (1): 16–30. Accessed September 28, 2020. doi:org/10.17113/ftb.56.01.18.5491.

[43] Hafsan, Hafsan, Nurhikmah Nurhikmah, Yuniar Harviyanti, Eka Sukmawati, Isna Rasdianah, Cut Muthiadin, Laily Agustina, Asmuddin Natsir, and Ahyar Ahmad. 2018. "The Potential of Endophyte Bacteria Isolated from *Zea Mays* L. as Phytase Producers." *Journal of Pure and Applied Microbiology* 12 (3): 1277–80. Accessed September 28, 2020. doi:org/10.22207/JPAM.12.3.29.

[44] Zhang, Shengpeng, Shao-an Liao, Xiaoyuan Yu, Hongwu Lu, Jian-an

Xian, Hui Guo, Anli Wang, and Jian Xie. 2015. "Microbial Diversity of Mangrove Sediment in Shenzhen Bay and Gene Cloning, Characterization of an Isolated Phytase-Producing Strain of SPC09 B. Cereus." *Applied Microbiology and Biotechnology* 99 (12): 5339–50. https://doi.org/10.1007/s00253-015-6405-8.

[45] Hussin, Anis Shobirin Meor, Abd-ElAziem Farouk, Ralf Greiner, Hamzah Mohd Salleh, and Ahmad Faris Ismail. 2007. "Phytate-Degrading Enzyme Production by Bacteria Isolated from Malaysian Soil." *World Journal of Microbiology and Biotechnology* 23 (12): 1653–60. Accessed September 28, 2020. doi:org/10.1007/s11274-007-9412-9.

[46] Konietzny, Ursula, and Ralf Greiner. 2004. "Bacterial Phytase: Potential Application, in Vivo Function and Regulation of Its Synthesis." *Brazilian Journal of Microbiology* 35 (1–2): 12–18. Accessed September 28, 2020. doi:org/10.1590/S1517-83822004000100002.

[47] Jorquera, Milko A., Stefanie Gabler, Nitza G. Inostroza, Jacquelinne J. Acuña, Marco A. Campos, Daniel Menezes-Blackburn, and Ralf Greiner. 2018. "Screening and Characterization of Phytases from Bacteria Isolated from Chilean Hydrothermal Environments." *Microbial Ecology* 75 (2): 387–99. Accessed September 28, 2020. doi:org/10.1007/s00248-017-1057-0.

[48] Aziz, G., M. Nawaz, A. A. Anjum, T. Yaqub, Mansur Ud Din Ahmed, J. Nazir, S. U. Khan, and K. Aziz. 2015. "Isolation and Characterization of Phytase Producing Bacterial Isolates from Soil." *Journal of Animal and Plant Sciences*. 25(3): 771-776.

[49] El-Toukhy, Nabil M. K., Amany S. Youssef, and Mariam G. M. Mikhail. 2013. "Isolation, Purification and Characterization of Phytase from *Bacillus Subtilis* MJA." *African Journal of Biotechnology* 12 (20): 2957–67. Accessed July 21, 2020. doi:org/10.5897/AJB2013.12304.

[50] Radu, Son, and Cheah Yoke Kqueen. 2002. "Preliminary Screening of Endophytic Fungi from Medicinal Plants in Malaysia for Antimicrobial and Antitumor Activity." *The Malaysian Journal of*

Medical Sciences : MJMS 9 (2): 23–33. doi://www.ncbi.nlm.nih.gov/pubmed/22844221.

[51] Bae, H. D., L. J. Yanke, K. J. Cheng, and L. B. Selinger. 1999. "A Novel Staining Method for Detecting Phytase Activity." *Journal of Microbiological Methods.* Accessed July 21, 2020. doi:org/10.1016/S0167-7012(99)00096-2.

[52] McInroy, John A., and Joseph W. Kloepper. 1995. "Survey of Indigenous Bacterial Endophytes from Cotton and Sweet Corn." *Plant and Soil* 173 (2): 337–42. Accessed July 21, 2020. doi:org/10.1007/BF00011472.

[53] Shobirin, Anis, Meor Hussin, Abd-elaziem Farouk, Abdul Manaf Ali, and Ralf Greiner. 2010. "Production of Phytate-Degrading Enzyme from Malaysian Soil Bacteria Using Rice Bran Containing Media." *Journal of Agrobiotechnology* 1 (1): 17–28.

[54] Magnani, G. S., C. M. Didonet, L. M. Cruz, C. F. Picheth, F. O. Pedrosa, and E. M. Souza. 2010. "Diversity of Endophytic Bacteria in Brazilian Sugarcane." *Genetics and Molecular Research* 9 (1): 250–58. Accessed July 21, 2020. doi:org/10.4238/vol9-1gmr703.

[55] Vohra, Ashima, and T. Satyanarayana. 2003. "Phytases: Microbial Sources, Production, Purification, and Potential Biotechnological Applications." *Critical Reviews in Biotechnology* 23 (1): 29–60. Accessed September 28, 2020. doi:org/10.1080/713609297.

[56] Pandey, Ashok, George Szakacs, Carlos R. Soccol, Jose A. Rodriguez-Leon, and Vanete T. Soccol. 2001. "Production, Purification and Properties of Microbial Phytases." *Bioresource Technology.* Accessed September 28, 2020. doi:org/10.1016/S0960-8524(00)00139-5.

[57] Kusumadjaja, Aline Puspita, Tutuk Budiati, Sajidan Sajidan, and Ni Nyoman Tri Puspaningsih. 2010. "Karakterisasi Ekstrak Kasar Fitase Termofilik Dari Bakteri Kawah Ijen Banyuwangi, Isolat AP-17." *Berkala Penelitian Hayati* 16 (1): 9–14. Accessed Oktober 16, 2020. doi:org/10.23869/bphjbr.16.1.20102. [Characterization of Thermophilic Phytase Crude Extract from Bacteria Ijen Banyuwangi Crater, Isolate AP-17. *Berkala Penelitian Hayati* 16 (1): 9–14.].

[58] Brock, Thomas D, Michael T Madigan, and John M Martinko. 2006. "Biology of Microorganisms, Eleventh Edition." *Biology of Microorganisms, Eleventh Edition*. Pearson; Prentice Hal.

[59] Patantis, Gintung, and Yusro Nuri Fawzya. 2009. "Teknik Identifikasi Mikroorganisme Secara Molekuler." *Squalen Bulletin of Marine and Fisheries Postharvest and Biotechnology* 4 (2): 72. Accessed Oktober 16, 2020. doi:org/10.15578/squalen.v4i2.146.

[60] Sogandi. 2018. *Biologi Molekuler: Identifikasi Bakteri Secara Molekuler*. 1st ed. Jakarta: Universitas 17 Agustus 1945 Press. [*Molecular Biology: Molecular Identification of Bacteria*. 1st ed. Jakarta: Universitas 17 Agustus 1945 Press.].

[61] Unno, Yusuke, Kenzo Okubo, Jun Wasaki, Takuro Shinano, and Mitsuru Osaki. 2005. "Plant Growth Promotion Abilities and Microscale Bacterial Dynamics in the Rhizosphere of Lupin Analysed by Phytate Utilization Ability." *Environmental Microbiology* 7 (3): 396–404. Accessed Oktober 16, 2020. doi:org/10.1111/j.1462-2920. 2004.00701.x.

[62] Bontemps, Cyril, Geoffrey N. Elliott, Marcelo F. Simon, Fábio B. Dos Reis Júnior, Eduardo Gross, Rebecca C. Lawton, Nicolau Elias Neto. 2010. "Burkholderia Species Are Ancient Symbionts of Legumes." *Molecular Ecology* 19 (1): 44–52. Accessed Oktober 16, 2020. doi:org/10.1111/j.1365-294X.2009.04458.x.

[63] Rhodes, Katherine A., and Herbert P. Schweizer. 2016. "Antibiotic Resistance in Burkholderia Species." *Drug Resistance Updates* 28 (September): 82–90. Accessed Oktober 16, 2020. doi:org/10.1016/j. drup.2016.07.003.

[64] Vanlaere, E., John J. LiPuma, Adam Baldwin, Deborah Henry, E. De Brandt, Eshwar Mahenthiralingam, David Speert, Chris Dowson, and Peter Vandamme. 2008. "*Burkholderia Latens Sp. Nov., Burkholderia Diffusa Sp. Nov., Burkholderia Arboris Sp. Nov., Burkholderia Seminalis Sp. Nov. and Burkholderia Metallica Sp. Nov.*, Novel Species within the *Burkholderia Cepacia Complex*." *International Journal of Systematic and Evolutionary Microbiology* 58 (7): 1580–90. Accessed Oktober 16, 2020. doi:org/10.1099/ijs.0.65634-0.

[65] Graminho, Eduardo Rezende, Naoki Takaya, Akira Nakamura, and Takayuki Hoshino. 2015. "Purification, Biochemical Characterization, and Genetic Cloning of the Phytase Produced by Burkholderia Sp. Strain A13." *Journal of General and Applied Microbiology.* Accessed Oktober 16, 2020. doi:org/10.2323/jgam.61.15.

[66] Suárez-Moreno, Zulma Rocío, Jesús Caballero-Mellado, Bruna G. Coutinho, Lucia Mendonça-Previato, Euan K. James, and Vittorio Venturi. 2012. "Common Features of Environmental and Potentially Beneficial Plant-Associated *Burkholderia*." *Microbial Ecology* 63 (2): 249–66. Accessed Oktober 16, 2020. doi:org/10.1007/s00248-011-9929-1.

[67] Yaish, Mahmoud W., Irin Antony, and Bernard R. Glick. 2015. "Isolation and Characterization of Endophytic Plant Growth-Promoting Bacteria from Date Palm Tree (*Phoenix Dactylifera* L.) and Their Potential Role in Salinity Tolerance." *Antonie van Leeuwenhoek* 107 (6): 1519–32. Accessed Oktober 16, 2020. doi:org/10.1007/s10482-015-0445-z.

[68] Kryuchkova, Yelena V., Gennady L. Burygin, Natalia E. Gogoleva, Yuri V. Gogolev, Marina P. Chernyshova, Oleg E. Makarov, Evgenii E. Fedorov, and Olga V. Turkovskaya. 2014. "Isolation and Characterization of a Glyphosate-Degrading Rhizosphere Strain, *Enterobacter Cloacae* K7." *Microbiological Research* 169 (1): 99–105. Accessed Oktober 16, 2020. doi:org/10.1016/j.micres.2013.03.002.

[69] Kalsi, Harpreet Kaur, Rajveer Singh, Harcharan Singh Dhaliwal, and Vinod Kumar. 2016. "Phytases from Enterobacter and Serratia Species with Desirable Characteristics for Food and Feed Applications." *3 Biotech* 6 (1): 64. Accessed Oktober 16, 2020. doi:org/10.1007/s13205-016-0378-x.

[70] Chanderman, Ashira, Adarsh Kumar Puri, Kugen Permaul, and Suren Singh. 2016. "Production, Characteristics and Applications of Phytase from a Rhizosphere Isolated *Enterobacter* Sp. ACSS." *Bioprocess and Biosystems Engineering* 39 (10): 1577–87. Accessed Oktober 16, 2020. doi:org/10.1007/s00449-016-1632-7.

[71] Yoon, Seong Jun, Yun Jaie Choi, Hae Ki Min, Kwang Keun Cho, Jin Wook Kim, Sang Cheol Lee, and Yeon Hoo Jung. 1996. "Isolation and Identification of Phytase-Producing Bacterium, Enterobacter Sp. 4, and Enzymatic Properties of Phytase Enzyme." *Enzyme and Microbial Technology* 18 (6): 449–54. Accessed July 16, 2020. doi:org/10.1016/0141-0229(95)00131-X.

[72] Taghavi, Safiyh, and Daniel van der Lelie. 2013. "Genome Sequence of the Plant Growth-Promoting Endophytic Bacterium Enterobacter Sp. 638." In *Molecular Microbial Ecology of the Rhizosphere*, 899–908. Hoboken, NJ, USA: John Wiley & Sons, Inc. Accessed July 16, 2020. doi:org/10.1002/9781118297674.ch84.

[73] Shoebitz, Mauricio, Claudia M. Ribaudo, Martín A. Pardo, María L. Cantore, Luigi Ciampi, and José A. Curá. 2009. "Plant Growth Promoting Properties of a Strain of Enterobacter Ludwigii Isolated from Lolium Perenne Rhizosphere." *Soil Biology and Biochemistry* 41 (9): 1768–74. Accessed Oktober 16, 2020. doi:org/10.1016/j.soilbio.2007.12.031.

[74] Koryagina, A. O., D. S. Bul'makova, A. D. Suleimanova, N. L. Rudakova, A. M. Mardanova, S. Y. Smolencev, and M. R. Sharipova. 2019. "Bacterial Enzymes as Potential Feed Additives in Poultry Farming." *Uchenye Zapiski Kazanskogo Universiteta. Seriya Estestvennye Nauki* 161 (3): 459–71. Accessed Oktober 16, 2020. doi:org/10.26907/2542-064X.2019.3.459-471.

[75] Greiner, Ralf, and Sajidan. 2008. "Production of D-Myo-Inositol (1,2,4,5,6) Pentakisphosphate Using Alginate-Entrapped Recombinant *Pantoea Agglomerans* Glucose-1-Phosphatase." *Brazilian Archives of Biology and Technology* 51 (2): 235–46. Accessed Oktober 16, 2020. doi:org/10.1590/S1516-89132008000200002.

[76] Greiner, Ralf. 2004. "Purification and Properties of a Phytate-Degrading Enzyme from Pantoea Agglomerans." *Protein Journal*. Accessed Juli 22, 2020. doi:org/10.1007/s10930-004-7883-1.

[77] Suleimanova, Aliya D., Anna A. Toymentseva, Eugenia A. Boulygina, Sergey V. Kazakov, Ayslu M. Mardanova, Nelly P. Balaban, and Margarita R. Sharipova. 2015. "High-Quality Draft

Genome Sequence of a New Phytase-Producing Microorganism Pantoea Sp. 3.5.1." *Standards in Genomic Sciences* 10 (1): 95. Accessed September 16, 2020. doi:org/10.1186/s40793-015-0093-y.

[78] Hafsan, Nurjannah, Cut Muthiadin, Isna Rasdianah, Ahyar Ahmad, Laily Agustina, and Asmuddin Natsir. 2018. "Phytate Activity of Thermophilic Bacteria from Sulili Hot Springs In Pinrang District South Sulawesi." *Scripta Biologica*, no. 36: 1–4. Accessed July 16, 2020. doi:org/10.20884/1.sb.2018.5.3.819.

[79] Wulandari, Rita. 2011. "Analisis Gen 16S rRNA Pada Bakteri Penghasil Enzim Fitase." Tesis, Universitas Sebelas Maret Surakarta, Indonesia [Analysis of 16S rRNA Gene in Phytase Enzyme Producing Bacteria. Thesis, Sebelas Maret University Surakarta, Indonesia].

[80] Sari, Evy Novita. 2012. *Identifikasi Bakteri Penghasil Fitase Berdasarkan Gen 16s Rrna Dan Karakterisasi Fitase Dari Kawah Sikidang Dieng.* Universitas Sebelas Maret Surakarta. [*Identification of Phytase Producing Bacteria Based on 16S rRNA Genes and Characterization of Phytase From Sikidang Dieng Crater.* Sebelas Maret University Surakarta].

[81] Hafsan, L. Agustina, A. Natsir, and A. Ahmad. 2020. "The Stability of Phytase Activity from *Burkholderia* Sp. Strain HF.7." *EurAsian Journal of BioSciences* 14 (1): 973–76.

[82] Scopes, Robert K. 2002. "Enzyme Activity and Assays." In *Encyclopedia of Life Sciences*. Chichester: John Wiley & Sons, Ltd.

[83] Brown, Susan. 2008. "Enzyme Activity." In *IUPAC Compendium of Chemical Terminology*. Research Triangle Park, NC: IUPAC.

[84] Daniel, R. M., R. V. Dunn, J. L. Finney, and J. C. Smith. 2003. "The Role of Dynamics in Enzyme Activity." *Annual Review of Biophysics and Biomolecular Structure* 32 (1): 69–92. Accessed Augustus 9, 2020.doi:org/10.1146/annurev.biophys.32.110601.142445.

[85] Gauthier, Robert. 2002. "Intestinal Health, the Key to Productivity." In *Nutrition*, 332.

[86] Shimizu, Mikio. 1992. "Purification and Characterization of Phytase from Bacillus Suhtilis (Natto) N–77." *Bioscience, Biotechnology, and Biochemistry* 56 (8): 1266–69. Accessed Augustus 9, 2020.

doi:org/10.1271/bbb.56.1266.

[87] Buddrick, Oliver, Oliver A. H. Jones, Hugh J. Cornell, and Darryl M. Small. 2014. "The Influence of Fermentation Processes and Cereal Grains in Wholegrain Bread on Reducing Phytate Content." *Journal of Cereal Science* 59 (1): 3–8. Accessed Augustus 9, 2020. doi:org/10.1016/j.jcs.2013.11.006.

[88] Ishiguro, Takahiro, Tomotada Ono, Katsuhiko Nakasato, Chigen Tsukamoto, and Shinji Shimada. 2003. "Rapid Measurement of Phytate in Raw Soymilk by Mid-Infrared Spectroscopy." *Bioscience, Biotechnology, and Biochemistry* 67 (4): 752–57. Accessed Augustus 9, 2020. doi:org/10.1271/bbb.67.752.

BIOGRAPHICAL SKETCH

Hafsan

Affiliation: Universitas Islam Negeri Alauddin, Indonesia

Education: Doctoral Program in agrobiotechnology Hasanuddin University Makassar, Indonesia

Research and Professional Experience: Biotechnology, Microbiology, Agrobiotechnology, Enzymology

Professional Appointments: Associate professor of biotechnology; executive chairman of the quality assurance committee.

Honors:

1. The best researcher of the national development cluster in 2019;
2. Awarde satya lencana 10 years of devotion in 2020.

Publications from the Last 3 Years:

1. Hafsan, Laily Agustina, Asamuddin Natsir, Ahyar Ahmad. 2020. The stability of Phytase activity from Burkholderia sp. strain HF.7. *Eurasian Journal of Biosciences* 14 (1), 991-994;
2. Hafsan, Fatmawati Nur, Muhammad Halifah Mustami, Khaerani Kiramang, Rahmaniah. 2020. Functional Characteristics of Lactobacillus Fermentum Origin of Whey (Waste Processing) Dangke Products as Probiotic Candidate. *International Journal of Current Science and Multidisciplinary Research* 3 (8), 236-241;
3. Aziz, IR, C Muthiadin, Hafsan. 2019. Biodegradasi Plastik LDPE Hitam Dan Putih Pada Sampah TPA Antang dalam Kolom Winogradsky. Al-Kauniyah: *Jurnal Biologi* 12 (2), 164-170;
4. Hafsan, M Maslan, M Masri, L Agustina, A Natsir, A Ahmad. 2019. Pengaruh Variasi Media terhadap Aktivitas Fitase Burkholderia Sp. Strain HF. 7. *Bionature* 20 (1) 112-119.
5. Hafsan, S Sukmawaty, M Masri, IR Aziz, SL Wulandari. 2018. Antioxidant Activities of Ethyl Acetic Extract of Endophytic Fungi from Caesalpinia sappan L. and Eucheuma sp. International Journal of Pharmaceutical Research 11 (1), 1-6;
6. Hafsan, N Nurhikmah, Y Harviyanti, E Sukmawati, I Rasdianah. 2018. The Potential of Endophyte Bacteria Isolated from *Zea mays* L. as Phytase Producers. *Journal of Pure and Applied Microbiology* 12 (3), 1-4.
7. Hafsan, Eka sukmawaty, Mashuri Masri, Ahyar Ahmad, Laily Agustina and Asmuddin Natsir. 2018. Phytase Activity of Four Endophytes Bacteria from *Zea Mays* L. *11th International Conference on Chemical, Agricultural, Biological and Environmental Sciences* (CABES-2018) ISBN 978-93-84422-85-1;
8. Hafsan, N Nurjannah, C Muthiadin, IR Azis, A Ahmad, L Agustina. 2018. Phytate Activity of Thermophilic Bacteria from Sulili Hot Springs in Pinrang District South Sulawesi. *Scripta Biologica* 5 (3).

9. Hafsan, Y Anriana, F Nur, K Kiramang, MK Mustami. 2017. Potency of Bacteriocin by Pediacoccus acidalacticy Indigenous Dangke as Biopreservative for Meatballs. *Proceedings 5th National & International-HCU Thailand.*
10. Arfani, N, F Nur, Hafsan, R Azrianingsih. 2017. Bacteriocin production of Lactobacillus sp. from intestines of ducks (Anas domesticus L.) incubated at room temperature and antibacterial effectivity against pathogen. *AIP Conference Proceedings* 1844 (1), 030004;
11. Hafsan, II Irwan, L Agustina, A Natsir, A Ahmad. 2017. Isolation and characterization of phytase-producing thermophilic bacteria from Sulili Hot Springs in South Sulawesi. *Scientific Research Journal* (SCIRJ) 5 (12), 1-4;

In: *Zea mays* L.: Cultivation, and Uses
Editor: Sarah Dunn
ISBN: 978-1-53619-181-3
© 2021 Nova Science Publishers, Inc.

Chapter 2

MAIZE (*ZEA MAYS* L.) SUITABILITY FOR WET MILLING AND ANIMAL NUTRITION IN RELATION TO PHYSICAL AND CHEMICAL QUALITY PARAMETERS

Marija Milašinović-Šeremešić[1,*], *Olivera Đuragić*[1], *Milica Radosavljević*[2] *and Ljubica Dokić*[3]

[1]Research Center for Feed Technology and Animal Products, Institute of Food Technology in Novi Sad, University of Novi Sad, Novi Sad, Serbia
[2]Department of Food Technology and Biochemistry, Maize Research Institute "Zemun Polje," Belgrade, Serbia
[3]Department of Carbohydrate Food Engineering, Faculty of Technology, University of Novi Sad, Novi Sad, Serbia

ABSTRACT

Maize (*Zea Mays* L.), being very competitive as a high carbohydrate yielding plant, represents the most important raw material for the commercial starch production (wet-milling industry) and a major energy feed ingredient in almost all animal diets. Maize kernel and whole plant varies in compositional traits and digestibility due to genetics and numerous environmental factors.

The focus of this study is on the physical traits, chemical composition, and their relationship with wet-milling properties and nutritional quality parameters of maize hybrids of different maturity groups and various endosperm types (dent, semi-dent and flint). The selected hybrids were grown under the semiarid conditions with the application of the same cropping practices at the experimental field of the Maize Research Institute in Zemun Polje, Belgrade, Serbia. Furthermore, the aim was to characterize suitability of the maize hybrids for wet-milling and animal nutrition.

Through many years of our research of maize quality and utilization, the obtained results showed significant relationship between quality parameters and contributed in classifying their importance and relevance. Accordingly, obtained relationships enabled the estimation and prediction of maize quality (utility value) for a particular purpose. A significant negative correlation was found between kernel protein content and portion of soft endosperm as well as a significant positive correlation between kernel protein content and two physical parameters, milling response and density. Among the chemical composition parameters only starch content significantly affected the starch yield and negatively affected the gluten yield. Physical parameters of the kernel such as test weight, kernel density and hardness significantly affected starch yield and recovery. Hybrids with a lower test weight and density and a greater proportion of soft endosperm fraction had a higher yield, recovery and purity of starch.

A significant negative correlation was determined between the NDF (Neutral Detergent Fibres) content and the whole plant dry matter digestibility (IVDMD), as well as, between the hemicellulose content and the digestibility. Based on the results, it can be concluded that a maize hybrid, intended for a high value animal feed, should have a low content in ADL/NDF (ADL-lignin NDF ratio, %), because it negatively influences whole plant digestibility. Furthermore, a very significant positive correlation was also found among all assayed lignocellulose fibers components.

The study demonstrates the importance of evaluating both the chemical and physical quality parameters of the maize kernels due to screen and estimate the suitability (utility value) of maize kernel for wet milling and animal nutrition. Understanding how maize plant cell wall constituents affect IVDMD is an important goal of future breeding research programs in order to improve forage utilization in animal feeding. The results obtained in our studies indicate that genetic differences in the long-term of maize hybrids breeding programs development can lead to providing farmers and industry with hybrids of good quality, desirable properties, and acceptable yield under the variable climatic conditions and with a lower cost.

Keywords: maize, kernel, whole plant, wet milling, animal nutrition, physical traits, chemical composition

INTRODUCTION

Based on the cultivated areas and produced quantities, maize is a major field crop in Serbia. Most of maize production in Serbia is rainfed with variable management inputs and hybrid that range from FAO 100-800. The greatest part of maize produced in both, our country and the world, is used as processed or unprocessed product in domestic animal feeding. Silage is one of the most important forms of maize utilization (Milašinović et al., 2017). In Serbia, out of total areas (1.2 million ha) sown with mercantile maize, approximately 5% are used for silage. Therefore, the development of new silage maize hybrids and studies on the effect of maize silage on intake, digestion, ruminal fermentation, lactation performance, and milk quality of dairy cows have been increasingly gaining in importance worldwide (Thomas et al., 2001; Ferraretto and Shaver; 2015; Khan et al., 2015). Many agroeconomic factors contribute to the fact that maize is a forage crop with no competitors in a large number of countries in the world. Among numerous factors, the following ones are the most important: possibility of obtaining high and stable yields by selection of hybrids greatly adaptable to specific conditions of each region, quality of maize biomass with a great proportion of fermentable carbohydrates and costs, i.e., the price of nutritive units achieved per hectare, are usually significantly lower than costs of achieved nutritive units of other crops (Radosavljević et al., 2012). The silage maize hybrids are generally characterized with high yield potential, valuable chemical components and good digestibility (Hegyi et al., 2009). The highest yielding grain hybrids were not necessarily the highest yielding silage hybrids. Silage quality of dent maize has reported in the literature to range from 54 to 86% dry matter digestibility, 7-11.5% protein content, 23-43% ADF and 40-68% NDF (Lauer et al., 2001). According to the study carried out by the same author in 2011 the range for NDF and digestibility among commercial maize hybrids is relatively narrow. Furthermore, the yield and quality differences among hybrids have increased and these trends are continuing.

Maize grain provides more feed for livestock than any other cereal grain (Maize Annual Report, 2018). Maize kernel is primary source of energy for domestic animal nutrition. Starch is its major nutritional and energetic component providing up to 68 to 74 percent of the kernel weight.Starch is a carbohydrate component that has the greatest influence on maize grain yields. Moreover, starch is a very important raw material in making numerous diverse products and bioethanol as a renewable alternative energy source.

In order to access quantity and fully define the carbohydrate content of maize, as well as, its nutritive value, it is also necessary to study the structure of cell walls of the whole plant. All carbohydrates in plant nutrients are grouped into: 1) structural carbohydrates (carbohydrates of cell walls), which include NDF (neutral detergent fibres - hemicelluloses + cellulose + lignin), ADF (acid detergent fibres – cellulose + lignin), ADL (lignin) and 2) non-structural carbohydrates - NFC (carbohydrates present in the plant cell content) that are made of starch, sugars and pectin. Many authors have reported that differences in the genetic background of maize genotypes affected the chemical composition, especially the ADF, NDF, starch and protein content (Thomson et al., 2001; Johnson et al., 2002; Schwab et al., 2003; Krakowski, 2006; Radosavljević et al., 2012) and the dry matter content (Szyskowska et al., 2007).

The maize kernel is also an important source of dietary proteins. However, compared to legume seeds, its nu-tritional quality is poor due to deficiency of two essential amino acids, lysine and tryptophan. Although the germ protein has adequate lysine content (5.4%) in whole kernel, this is diluted by the much more abundant endosperm proteins, which have an average lysine content of only about 1.9%. This is because 60–70% of endo-sperm protein consists of zeins, which contain few or no lysine residues (Coleman and Larkins, 1999). Similarly, the absence of tryptophan residues in zein proteins is the reason for the low tryptophan content of maize protein.However, amino acid contents in maize endosperm can be improved by mutant selection (Muehlbauer et al., 1994) or genetic engineering (Huang et al., 2006). Breeding for quality protein maize would have the added advantage of biofortification of maize. Wild relatives have been regarded as a source to extend the genetic diversity of maize breeding programs.

Maize kernel is also a source of oil which is highly regarded for human consumption as it reduces the blood cholesterol concentration (Dupont et al., 1990). Chemical composition in maize kernel is genetically controlled, and the presence of genetic di-versity is essential for maize quality and utilization improvement (Radosavljević et al., 2010). In addition, physical kernel traits may have an effect on nutritive value of maize. Previous research indicated that maize hardness is in relationship to physical kernel traits and subsequently with ruminal starch availability (Correa et al., 2002), feed efficiency of feedlot cattle (Jaeger et al., 2006), and growth performance and carcass characteristics of pigs (More et al., 2008).

Maize, as we usually think of it, is primarily feed and food grain. However, there are many uses and applications of maize. The fastest growing are industrial uses. Maize is processed by three major industries: wet millers that produce starch, sweeteners and maize oil; dry millers that produce maize flour and grits; and distillers that produce beverage and bioethanol.

Renewability of maize as a raw material and growing environmental pollution by oil products represent two principal reasons for maize becoming one of the major feedstocks for the energy production. Alternative fuel-bioethanol is mostly produced from starchy parts of the maize grain leaving significant amounts of valuable by-products such as distillers' dried grains with solubles (DDGS), which can be used as a substitute for traditional feedstuff (Semenčenko et al., 2014).

Maize wet millers prefer hybrids with higher portion of soft endosperm and higher starch content which are easier to steep for a shorter time and later on, this provides more effective starch and gluten separation. Production of the starches and ethanol are now most important technological maize uses, but accessibility and low cost of maize gluten could bring to its wider application (Šeremešić et al., 2016). Experience indicates that 70% of starch yield variability is caused by genetics and 30% by environmental conditions (Eckhoff, 1999). Zehr et al. (1995, 1996) found that wet-milling properties, including starch yield, are heritable.The contemporarycommercial maize starch production is based on the concept of the traditional maize starch processing, i.e., on wet milling. Maize starch as an essential product of primary starch processing, represents an initial raw material for numerous technological and biotechnological processes in further industrial reproduction, i.e., in higher stages of processing. At present, there are a series of products made on the basis of maize starch (Maize Annual Report, 2018).

In wet-milling process the embryos, which are rich in lipid, are isolated and a product known as maize (corn) germ oil is produced by extraction. By-products of maize wet milling, germ (oil) and gluten meal (protein source) are very suitable for inclusion in feed.

As the market increases for specialty maize hybrids, maize genetically bred with unique starch properties or maize hybrid with altered chemical composition, a thorough scientific understanding of steeping chemistry and the entire wet milling process will become increasingly important.

Regarding to various physical traits, nutritional quality parameters and process characteristics (wet-milling properties) the objectives of this study was to characterize differences and suitability of the selected Serbian maize genotypes for animal nutrition and wet-milling. The data was correlated to find the interrelationship between these parameters in order to predict the maize quality and suitability for certain uses

METHODS

ZP hybrids of the FAO maturity groups 100-800 were analysed. The two-replicate trail was set up according to the randomised complete-block design in the experimental field of the Maize Research Institute. The experimental plot size amounted to 21m2, while sowing density was

60,000 plants ha-1. Plants of each replicate were harvested in the full waxy maturity stage from the area of 7m2 (two inner rows), and yields of fresh biomass of the whole plants, plants without ears and ears were estimated. Five average plants per replicate were selected for further tests. Samples of the whole plants were cut and dried at 60oC for 48h. In order to determine the content of dry matter, the whole plant samples were ground in the 1-mm mesh mill. Then, theanalysis of the absolute dry matter was done on the oven dry basis (105oC for 12 h) in order to estimate the total dry matter. Moreover, the analysis of the content of forage fibres (NDF-Neutral Detergent Fibres, ADF-Acid Detergent Fibres, ADL-Acid Detergent Lignin, hemicelluloses, cellulose) was performed by the Van Soest detergent method modified by Mertens (1992). In vitro digestibility of the whole maize plant was done by the Aufréré method (Aufréré, 2006), which is based on the hydrolysis of proteins of the whole plant in the pepsin acid solution (Merck 2000 FIP u/g Art 7190) at 40oC for 24 h, and then on the hydrolysis of carbohydrates in the cellulase solution (cellulose Onozuka R10) in duration of 24 h. The NDFD was calculated by the equation reported by Brenner et al. (2010): NDFD=100(ES-(100-NDF))/NDF); the L/NDF ratio of the investigated maize hybrids was determined according to Frey et al. (2004).

Figure 1. Experimental field in Zemun Polje (Maize Research Institute "Zemun Polje"), Serbia.

The 1000-kernel weight was evaluated by counting and weighting of 4×250 of un-broken kernels. Kernel density. Approximately 33 g of whole kernels was weighed to ±0.001 g. Volume determinations were then made with a Beckman model 930 air comparison pycnometer. Procedures for using the air comparison pycnometer are described in previous published paper (Radosavljević et al., 2010). The analyses were performed in three replicates.

The kernel hardness was measured by Stenvert-Pomeranz method by milling a 20 g of maize kernels in micro hammer-mill at 3600 rpm and 2-mm sieve (Pomeranz et al., 1985). Results were expressed as milling response and soft endosperm portion (%). The milling response presents the time (s) necessary for kernel grinding until the top level of the material collected in a glass cylinder (125 × 25 mm) reaches the level of 17 mL. The test was performed in three replicates.

The starch content was determined by Ewers polarimetric method (ISO, 1997). Amylose content has been determined by a rapid colorimetric method (Mc Grance et al., 1998). Dry matter content in the maize flour was determined by the standard drying method in an oven at 105 °C to constant mass. Oil concentration was determined accordingto the Soxhlet method (AOAC, 2000). Protein content was estimated as the total nitrogen by the Kjeldahl method multiplied by 6.25, and the ash content was determined by slow combustion of the sample at 650 °C for 2 h (AOAC, 1990). Crude fibre content was determined by Weende method adjusted for Fibretec™ Systems, Foss, Denmark (ISO, 1993). The amino acids analyses of maize kernel were performed using ion exchange chromatography with utilization of Automatic Amino Acid Analyzer Biochrom 30+ (Biochrom, Cambridge, UK), according to Spackman et al. (1958).

All chemical analyses were performed in three replications, and the results were statistically analyzed. A factorial analysis of variance (ANOVA) for trials was conducted using randomized complete block (RCB) design. Treatment means were tested using Tukey HSD test to determine the significant differences between group means in an analysis of variance setting at an alpha-level of 0.05. Pearson's product moment correlation coefficient was used for determining correlations between the estimated traits and Principal Component Analysis (PCA) to summarize the data of traits in fewer variables (the PC-axis or factors) and show which traits are close to each other, i.e., which carry comparable information. Hierarchical cluster analysis was used to group the hybrid into classes or clusters based on their similarities. All statistical analyses were done by the STATISTICA program package 13.3 (StatSoft Inc., 2018).

RESULTS AND DISCUSSION

Maize Grain for Wet-Milling (Starch Production)

The increasing interest by processing industry in maize as the important source of carbohydrate has resulted in broadening of breeding programmes on hybrids with specific traits and for special purposes (Milašinović-Šeremešić et al., 2018; 2019). Traditionally, the main goals of maize breeding are the production of high yielding hybrids tolerant to drought and pests. Little attention has been paid to the technological value (suitability and/or extractability) of maize grain for wet-milling (starch production). However, advances have been made by breeders within this area as well, resulting in maize kernels with a wide range of structures and compositions. By exploiting genetic variation, the composition of the kernel has been altered for both the quantity and quality of starch, proteins and oil throughout kernel development.

Compositional traits of maize are essential for various end-uses; feed for animals, food for humans, and raw materials for industry. The most important parameters for the estimation of the technological value and/or suitability of maize kernel in wet milling are as follows: yield, recovery and purity of starch, i.e., the protein content in isolated starch. High starch recovery and yields are the principal parameters of a well performed maize wet milling procedure. Wet-milling properties of each ZP maize hybrids were evaluated on the basis of yield (ration of extracted starch to amount of grain), recovery (ration of extracted starch to the starch content of grain), and purity of starch (protein content of starch). High starch yield, high starch recovery and low protein content in starch are indicators of good wet milling. The data for chemical composition, physical and wet-milling properties of 12 ZP maize hybrids are given in Table 1.

The obtained results show that starch yields of the studied maize hybrids ranged from 58.8% in ZP 633 to 69.0% in ZP 808, which correspond to the starch recovery of 83.7% and 93.7% respectively. The highest starch recovery (94.0%) was in the hybrid ZP 677 with the starch yield of 67.6%. The lowest (5.3%), i.e., the highest (13.5%) gluten yields were detected in ZP 808, i.e., ZP 633, respectively. Hybrids with a higher starch yield and recovery have a lower yield of gluten. The protein content in recovered starches ranged from 0.11 to 0.29% pointing out to high quality of produced starches. These data suggest that hybrids ZP 677 and ZP 808 are the most suitable for wet milling. Our results are in an agreement with Eckhoff's study (1999) which found that in 387 samples the starch yield ranged from 50% to 70%, with an average starch yield of 63.8% and a peak frequency of 64%.

The content of amylose in these starches is characteristic for the normal maize starches (approximately 24% of amylose and 76% of amylopectin) (data not presented). Viscoamylograms of the starches are also characteristic for the normal maize starches (pasting temperatures ranged from 72.5°C to 74.2°C, peak temperature from 80.5°C to 85.8°C, peak viscosity from 910BU to 1000BU, and set-back viscosity on 50°C from 1310BU to 1460BU) (data not presented).

On the basis of correlation analysis and gained correlation coefficients very high dependences are noticed between grain starch content and starch yield (0.76**), as well as the grain starch content and gluten yield (-0.71*). Starch content also positively affected the starch recovery (0.49) but it was not statistically significant. On this basis, it can be concluded that hybrids with the highest level of starch content in grain do not have to, as a rule, give the highest starch recovery. The grain starch content negatively affected the protein content in recovered starches which agreed with the results of Singh et al. (2001). The grain protein content affected positively but not statistically significant the gluten yield, and negatively the starch yield and recovery (Table 2).

Table 1. Chemical composition, physical and wet-milling properties of 12 ZP MAIZE hybrides[a,b]

Hybrid	Str	Pro	TWt	KWt	ADen	IF	SE	StrY	SRec	Glu	Germ	Fib	PiS
ZP 360	71.7	11.6	816.4	327.6	1.25	65.0	40.5	66.5	92.7	7.3	8.2	8.2	0.11
ZP 434	72.2	9.6	841.8	351.0	1.24	26.5	38.8	65.4	90.7	7.0	8.3	9.0	0.26
ZP 480	74.3	13.7	835.7	307.3	1.26	50.0	39.8	66.9	90.0	7.3	7.9	7.9	0.20
ZP 511	72.8	10.8	832.5	352.8	1.27	48.8	42.4	65.9	90.6	6.7	8.1	8.6	0.19
ZP 633	70.3	11.7	860.5	311.5	1.31	0.40	33.3	58.8	83.7	13.5	7.7	8.7	0.28
ZP 677	71.8	10.4	833.1	335.2	1.28	30.8	41.7	67.6	94.0	6.5	8.2	8.4	0.13
ZP 680	72.6	10.5	842.0	358.4	1.29	29.3	38.8	63.1	86.9	9.3	8.3	8.9	0.18
ZP 684	72.5	10.2	841.8	336.2	1.28	41.4	39.5	63.4	87.5	7.9	8.2	8.7	0.21
ZP 735	69.5	11.5	863.3	304.3	1.32	0.20	32.3	59.2	85.1	13.2	7.3	8.2	0.29
ZP 737	73.5	10.5	844.1	298.5	1.29	11.7	38.4	65.5	89.1	9.4	8.0	7.4	0.21
ZP 750	73.2	10.4	860.7	270.8	1.31	9.8	34.7	64.6	88.3	9.0	7.9	7.9	0.27
ZP 808	73.7	8.3	825.5	309.9	1.25	68.3	45.7	69.0	93.7	5.3	7.2	8.0	0.18
AVRG	72.34	10.77	841.45	321.96	1.28	31.85	38.83	64.66	89.36	8.53	7.94	8.33	0.21
SD	1.4	1.3	14.4	26.1	0.03	23.5	3.8	3.1	3.2	2.6	0.4	0.5	0.06

[a] Str = starch content (%), Pro = protein content (%), TWt = test weight (kgm-3), KWt = 1000-kernel weight (g), ADen = absolute density (gcm-3), IF = flotation index (%), SE = soft endosperm (%), StrY = starch yield (%), SRec = starch recovery (%), Glu = gluten yield (%), Germ = germ yield (%), Fib = fiber yield (%), PiS = protein in starch (%). [b] Calculated on a dry basis.

Table 2. Correlation coefficients between chemical composition, physical and wet-milling properties of various ZP maize hybrids[a]

	StrY	SRec	PiS	Glu	Germ	Fib
Str	0.76**	0.49	-0.34	-0.71*	0.20	-0.39
Pro	-0.29	-0.33	0.05	0.37	0.09	-0.15
TW$_t$	-0.80**	-0.84**	0.90**	0.82**	-0.29	0.08
KW$_t$	0.10	0.19	-0.40	-0.32	0.58	0.78**
ADen	-0.76**	-0.76**	0.55	0.82**	-0.29	-0.13
IF	0.09	0.04	-0.27	-0.32	0.18	0.48
SE	0.87**	0.85**	-0.76**	-0.91**	0.16	-0.03

[a] Str = starch content (%), Pro = protein content (%), TWt = test weight (kgm-3), KWt = 1000-kernel weight (g), ADen = absolute density (gcm-3), IF = flotation index (%), SE = soft endosperm (%), StrY = starch yield (%), SRec = starch recovery (%), Glu = gluten yield (%), Germ = germ yield (%), Fib = fiber yield (%), PiS = protein in starch (%).

The highest correlation coefficient was between soft endosperm portion and gluten yield (-0,91**). Starch yield and recovery as well as the protein content in recovered starches showed very high dependence to the soft endosperm portion (0.87**, 0.85** and -0.76**). Test weight shows very high negative correlation between starch yield and recovery (-0.80** and -0.84**), as well as a very high positive correlation between gluten yield and protein content in recovered starches (0.82** and 0.90**). Kernel density showed very high negative correlation to starch yield and recovery (-0.76** and -0.76**) while the interdependence of density and gluten yield was very high positive (0.82**). The same values of correlation coefficient (-0.76**) show that kernel density of the selected hybrids had the equal effect on starch yield and recovery. 1000-kernel weight showed very high effect on fiber yield (0.78**) (Table 2). On the basis of the above-mentioned relationships, it can be concluded that four quality parameters, test weight, kernel density, soft endosperm portion and grain starch content, had the highest effect on wet milling properties of the selected maize hybrids.

The results of regression analysis showed that there are statistically significant high regression dependence between the following pairs: starch yield-test weight (b=0.64**), starch yield-density (b=0.57**), starch yield-soft endosperm portion (b=0.76**), starch yield-flotation index (b=0.57**), starch yield-starch content (b=0.57**), starch recovery-test weight (b=0.70**), starch recovery-density (b=0.58**), starch recovery-soft endosperm portion (b=0.71**), starch recovery-flotation index (b=0.55**), protein content in recovered starches-test weight (b=0.80**) and protein content in recovered starches soft endosperm portion (b=0.58**). These regression relationships except the couples with flotation index are in good agreements to the obtained correlation dependences. Statistically significant high regression dependence was not found in the couple starch recovery-starch content in grain (b=0.24).

Test weight is a more easily measurable quality factor than kernel density and soft endosperm portion, the determinations which take much more time. Thus, this physical property is very suitable for predicting wet-milling properties.

Equations which can be used in prediction of wet-milling properties, i.e., starch yield, recovery and purity are as follows:

Starch yield (%) = -0.173 x TWt + 210.13, with b=0.64**
Starch yield (%) = 0.707 x SE + 37.22, with b=0.76**
Starch recovery (%) = -0.188 x TWt + 247.81, with b=0.70**
Starch recovery (%) = 0.713 x SE + 61.66, with b=0.71**

* and ** denote significance at 0.05 and 0.01 probability levels, respectively.

Using the multiple regression analysis, the following equation was developed:

Starch yield (%) = 65.83 + 1.22 x Str (%) - 0.106 x TWt,
with a multiple determination coefficient D = 0.78

Determination of starch yield is a very expensive, time consuming and complex procedure. The mathematical model developed in this study provides the prediction of starch yield on the basis of the test weight and grain starch content, parameters that are routinely determined.

MAIZE GRAIN FOR FEED

Maize is traditionally a feed grain and this continues to be an important means of use. The nutritional value of maize grain for feed is a function of the content of starch, oil, protein, fiber and antinutrients (primarily phytate, enzyme inhibitors and resistant starches).

Starch in cereals is the most abundant energy source for domestic animals. It is the main carbohydrate constituent of maize grain, makes two thirds of a maize grain dry matter (DM) and is, therefore, the most important economic component of maize grain (Milašinović-Šeremešić et al., 2019).

Typical chemical composition for the commodity yellow dent maize kernel on a dry matter basis is 71.7% starch, 9.5% protein, 4.3% oil, 1.4% ash, and 2.6% sugar (Watson, 2003).

The most recent results on grain quality (chemical composition and physical traits) testing of 55 ZP maize hybrids of various growing seasons and different endosperm types showed in Table 3 and 4.

Table 3. Chemical composition of 55 ZP hybrids[a]

	Starch (%)	Protein (%)	Oil (%)	Fiber (%)	Ash (%)
Min	53.00	8.04	4.87	1.90	1.16
Max	74.41	13.34	10.25	3.24	1.77
Average	67.67	9.85	6.33	2.45	1.36
SD	3.93	1.26	1.21	0.29	0.12

[a] Calculated on a dry basis.

Table 4. Physical traits of 55 ZP hybrids

Physical traits	KWt (g)	TWt (kgm^{-3})	Den (gcm^{-3})	IF (%)	MRes (s)	HE (%)	WAI	Pericarp (%)	Endosp. (%)	Germ (%)
Min.	104.1	590.0	1.16	0	7.4	42.6	0.20	5.1	71.1	8.4
Max	342.5	902.3	1.38	100	19.5	75.1	0.40	11.8	83.3	21.2
Average	249.1	764.4	1.25	58.63	10.6	58.4	0.25	6.9	80.6	12.6
SD	61.0	56.5	0.04	1.4	3.0	6.8	0.04	1.4	2.6	2.3

TWt = test weight, KWt = 1000-kernel weight, Den = absolute density, IF = flotation index, MRes = milling response, HE = hard endosperm, WAI = water absorption index.

These results, as well as previously achieved results on chemical traits of grain in dependence on the maize hybrid properties, pointed out that the greatest number of observed traits depending on the genetic base, i.e., on the type of hybrid, growing and environmental conditions, varied in a very wide range (Radosavljević et al., 2015). The variability of tested quality parameters provides wide possibilities for selecting maize hybrids as a raw material for certain purposes, as well as for selection of new hybrids (Milašinović-Šeremešić et al., 2019). Results obtained in this study are in accordance with results gained by foreign authors (Ketthaisong et al., 2015).

Out of 20 amino acids, an animal or a human being can synthesize only nine of them (non-essential amino acids). The remaining amino acids (essential amino acids-EAA) should be provided by their various sources of food. Arginine is re-graded as EAA in birds and fish. Therefore, it is regarded as a semi-essential amino acid. Cysteine and tyrosine are also re-graded as semi-essential amino acids, as they can be synthesized exclusively with methionine and phenylalanine, respectively (Boisen et al., 2000).

The amino acid profile of the selected maize genotypes demonstrated (data not presented) that the kernel is dominated with the amino acid glutamine (17.09-19.09%), followed by proline (8.44-10.03%) and leu-cine (9.54-11.65%). Lysine and methionine concentrations among the maize genotypes were very low (2.63-3.60% and 2.27-3.27%, respectively). The obtained results are in agreement with the results of the previous research (Wang et al., 2008). The observed differences between the genotypes in amino acids composition are not clear enough due to the fact that in this study has not considered other factors (environment, management practices etc.) that could affect the quality of kernel protein. Previous research has shown that nitrogen is an important macronutrient for the development of amino acids and proteins in maize crop (CIMMYT, 2003). Different field conditions altered the ratio of maize genetic effects and suppressed genetic effects for protein concentration (Kumar et al., 2018). The recent study demonstrated the nitrogen has a critical role in tryptophan and lysine production of quality protein maize (QPM) hybrid. On the other hand, the extensive research of Scott et al. (2006) suggests that the composition of maize hybrids has changed over time, while the quality of the protein (defined as methionine, lysine or tryptophan per protein) has not changed in a statistically detectable way. The authors found the kernel protein content of modern hybrids responds to plant density and environment differently than the protein content of older maize varieties. These differences may partially explain how modern hybrids can maintain yield in different environments, i.e., decrease of protein content in stressful environments frees resources that are used to maintain yield.

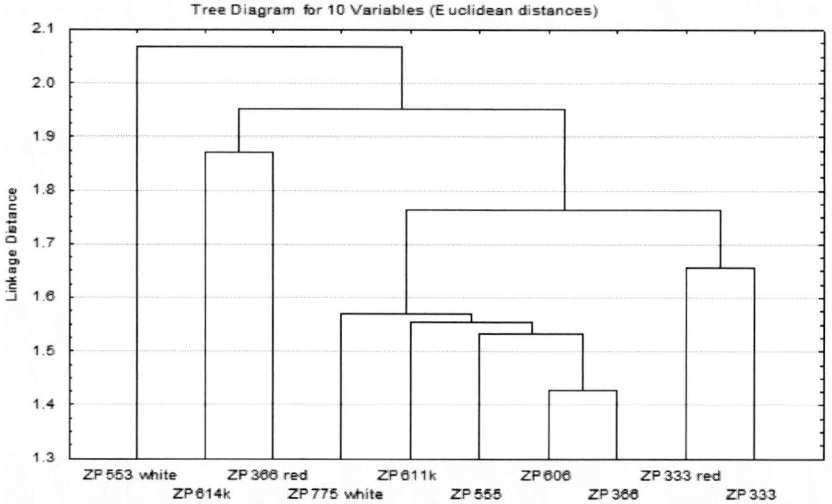

Figure 2. Dendrograms obtained from the hierarchical cluster analysis of each amino acid in the maize kernel.

Based on the above, it has not been observed a significant improvement in the amino acid composition regarding the specialty genotypes such as the selected white (ZP 553b, ZP 775b) and red (ZP 333c, ZP 366c) kernels and popping maize (ZP 611k, ZP 614k) genotypes.

The hierarchical cluster analysis (Figure 2) clearly shows four groups (clusters) of the genotypes which were differentiated on the basis of the similarity of amino acid profile. The genotypes were very good clustered according to their genetic background. The genotype ZP 553b has unrelated components with other genotypes. In their parental components, or maternal components, the genotypes ZP 366, ZP 606, ZP 555 and ZP 333 have a certain percentage of the genetic of the popcorn genotypes, therefore the popcorns are closely related to them. The genotypes ZP 333 and ZP 366 have the same maternal component. Obviously, there was no significant change in the amino acid composition of the red kernel of ZP 333c (compared with yellow kernel), as opposed to the genotype ZP 366, where the red kernel (ZP 366c) was grouped into the second subcluster. Relate to the white kernel genotype, ZP 775b, its maternal component contains a part of the germplasm that is close to the genetics of the genotypes ZP 606, ZP 555, ZP 333 and ZP 366.

Table 5. Pearson's product moment correlation coefficients between physical quality traits and chemical composition of different ZP maize genotypesa[a]

	KWt	Den	MRes	SE	Starch	Protein	Oil	Cellulose	Ash
KWt[b]	1.00	-0.92*	-0.96*	0.94*	0.60	-0.50	0.51	-0.55	0.12
Den[b]		1.00	0.96*	-0.98*	-0.63*	0.69*	-0.32	0.77*	-0.02
MRes[b]			1.00	-0.99*	-0.65*	0.67*	-0.51	0.62	-0.09
SE[b]				1.00	0.66*	-0.69*	0.45	-0.69*	0.04
Starch					1.00	-0.52	0.48	-0.31	-0.30
Protein						1.00	-0.34	0.66*	0.23
Oil							1.00	-0.11	-0.14
Cellulose								1.00	0.19
Ash									1.00

[a] Marked correlation are significant at p<0.05, N=10. [b] KWt - 1000-kernel weight; Den - density; MRes - milling response; SE - soft endosperm portion.

Both starch and protein contents affected hardness. On the basis of correlation analysis and gained correlation coefficients (Table 5) very high dependences are noticed between kernel starch content and the physical quality traits such as Den, MRes and SE (-0.63*, -0.65* and 0.66*). Protein content in maize kernel positively correlated with Den and MRes (0.69* and 0.67*), and negatively correlated with SE (-0.69*). Further, the kernel cellulose content affected positively on density and the protein content (0.77* and 0.66*) and negatively on SE (-0.69*).

On the basis of the mentioned relationships, it can be concluded that three physical quality traits, density, milling response and soft endosperm portion had the highest interdependence with the chemical quality parameters such as the kernel contents of starch, protein and cellulose of the selected materials. Maize genotypes with higher level of soft endosperm fraction in the whole kernel have higher starch content, and lower protein and cellulose content. Further, the increase in Den and MRes, resulted in the increase of protein content and the decrease of the starch content.

The results have indicated the significance of the physical parameters which are closely related to the nutritive quality and the utility value of maize kernel. The results agree with our previous findings (Milašinović et al., 2007; Semenčenko et al., 2013). Another important finding of this study is the negative (but not statistically significant) correlation between protein content and lysine content in kernel (-0.53) as well as the positive and significant correlation between protein content and methionine content (0.74*) (the data not shown).

On the PCA score plot, four groups of physical and chemical parameters are clearly separated. Compared to Pearson's product moment correlation PCA gives different perspectives over investigated traits. The correlation coefficients provide information about the strength of the association between two variables, while PCA provides orthogonal arrangement of variables and thus indicate their interrelation. Soft endosperm fraction, cellulose, density and protein are correlated on one side and negatively correlated with milling response and 1000-kernel weight on the other side. Starch and oil content are grouped, and closely correlated with milling response and 1000-kernel weight but opposite from ash content (Figure 3). The first principal component, PC1, accounts for 59.68% of the total data variance, and the second PC2 account for 15.07%. The separation of the investigated quality parameters is largely based on the first principal component.

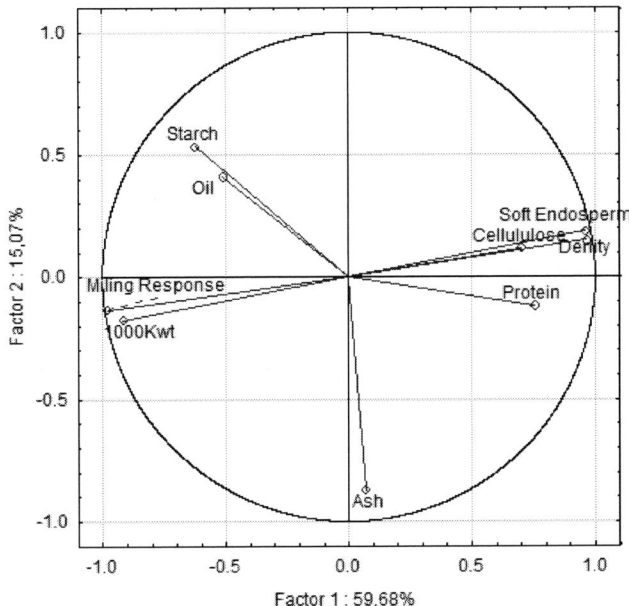

Figure 3. Principal component analyses loadings (similarities of 9 traits).

Previous research described the biochemical aspects of maize that have been related to maize hardness. Both starch and protein affect hardness, with most research focusing on the storage proteins (zeins). Both the content and composition of the zein fractions affect hardness (Žilić et al., 2011). Genotypes and growing environment influence the final protein and starch content and, to a lesser extent, composition. However, hardness is a highly heritable trait and, hence, when a desirable level of hardness is finally agreed upon, the breeders will quickly be able to produce material with the hardness levels required by the industry (Fox and Manley, 2009).

The observed interactions have to be examined in a larger number of maize genotypes. Further research regarding the inter-relationship among nutritional quality parameters of maize genotypes as well as the inter-relationship between physical quality traits and amino acids composition is necessary to establish stronger relationships.

MAIZE WHOLE PLANT FOR RUMINANT NUTRITION

Beside for grain production, maize is an important crop for forage production due to consistent quality and higher yield and energy contents than other forages, as well as because less labor and machinery time are needed for harvest and because costs per ton of dry matter are lower than of other harvested forages.

The plant cell walls are the most abundant source of carbohydrates in nature and represent the primary energy source for ruminant animals. The maize cell walls are predominantly composed of polysaccharides – cellulose and hemicellulose, and a highly complex polymer lignin (Lorenz, 2009). Therefore, forage quality assays based on the analyses of cell wall, i.e., fibers present in forages are of major concern in ruminant nutrition. The main quality parameters of maize biomass (whole plant) for silage are yield, structure of the dry matter yield, as well as the yield of the digestible dry matter. The silage maize hybrids are generally characterized with high yield potential, valuable chemical components and good digestibility.

In recent times, besides the content of lignocellulosic fractions, NDF digestibility, as a quality parameter of fibrous feeds, has been gaining in importance. Although NDF and ADF are good indicators of the plant fiber content in fibrous feeds, they do not show the degree of digestibility of these fibers. The NDF digestibility (NDFD) provides more precise data on the content of total nutrients (TDN), net energy content (NE) and a potential intake of fibrous feeds by animals. Generally, the increased NDF digestibility results in greater digestible energy and greater intake of fibrous feeds. Significant differences in dry matter digestibility and NDF digestibility of the whole plants were obtained in studies on the chemical composition of standard grain hybrids (Terzić et al., 2012). The dry matter digestibility and NDF digestibility of the whole plant of observed silage maize hybrids ranged from 57.02% to 66.16% and from 23.15% to 29.26%, respectively, while differences in dry matter digestibility and differences in NDF digestibility amounted to 9.14% and 6.11%, respectively (Terzić et al., 2012).

The results of digestibility of both dry matter and NDF of the whole plant of nine ZP hybrids of various genetic background and different FAO maturity groups that the NDF digestibility varied from 19.37% (ZP 505) to 31.86% (ZP 560), while the dry matter digestibility in observed ZP maize hybrids ranged from 59.67% (ZP 677) to 65.53% (ZP 434). Differences in NDF digestibility affected dry matter digestibility of the whole plant (Semenčenko et al., 2014). Some studies suggested that diary caws in lactations fed on fibrous feeds consumed greater amounts of dry matter and produced more milk if they were fed with fibrous feeds that contained more digestible NDF (Bertoia and Aulicino, 2014; Thomas et al., 2001).

The data presented in Tables 7 show the IVDMD, NDFD and the lignin to NDF ratio (L/NDF) of the whole plant of six recently studied maize hybrids. The lignin to NDF ratio (L/NDF) of whole plants ranged from 29 (ZP 666) to 39 g/kg (ZP 758 and ZP 802) and the IVDMD from 0.5667 (ZP 802) to 0.6734 (ZP 648), with the NDFD varying from 166 (ZP 758) to 322 g/kg (ZP 648).

Table 6. Whole plant lignocellulose fibers content of maize hybrids[a,b]

Hybrid	NDF (%)	ADF (%)	ADL (%)	Hemicelluloses (%)	Cellulose (%)
ZP 341	53.35[b]	26.32[a]	1.87[b]	27.03[b]	24.45[a]
ZP 427	48.07[d]	23.44[bc]	1.45[c]	24.63[d]	21.99[b]
ZP 648	48.14[d]	22.23[d]	1.45[c]	25.91[c]	20.78[d]
ZP 666	48.11[d]	22.67[cd]	1.42[c]	25.44[c]	21.25[cd]
ZP 758	49.21[c]	23.65[b]	1.93[b]	25.56[c]	21.32[bc]
ZP 802	56.89[a]	26.54[a]	2.21[a]	30.35[a]	24.33[a]
LSD $_{0.05}$	0.8	0.8	0.2	0.6	0.7

[a] NDF-Neutral Detergent Fibres, ADF-Acid Detergent Fibres, ADL-Acid Detergent Lignin. [b] Means in the same column with different superscripts differ (p<0.05).

The differences in the contents of NDF, ADF, ADL, hemicellulose, cellulose and IVDMD of the whole maize plant among observed ZP hybrids were 88.2, 43.1, 7.6, 57.2, 33.5 and 106.4 g/kg, respectively (Table 6). Obtained values for the content of lignocellulose fibers in the whole plant differed significantly among hybrids and were closely related to digestibility. The differences in the contents of lignocellulose fibers affected the differences in IVDMD. The hybrid ZP 802 had the highest content of NDF, ADF, ADL and hemicellulose and the lowest IVDMD coefficient (0.5667), while the hybrid ZP 648 had the lowest content of ADF and cellulose and the highest IVDMD coefficient (0.6734) of the whole plant for investigated hybrids. The same hybrid had the highest NDFD and the lowest lignin to NDF ratio of the whole maize plant as well.

Table 7. Whole plant IVDMD of maize hybrids[a,b]

Hybrid	Content (g kg^{-1})		Dry matter digestibility
	NDFD	LNDF^{-1}	(%)
ZP 341	309[a]	35[ab]	63.12[c]
ZP 427	319[a]	31[bc]	67.28[a]
ZP 648	322[a]	30[c]	67.34[a]
ZP 666	273[b]	29[c]	65.00[b]
ZP 758	166[d]	39[a]	58.96[d]
ZP 802	238[c]	39[a]	56.67[e]
LSD .05	16	5	0.8

[a] NDFD-digestibility of Neutral Detergent Fibres, LNDF-1-ADL-lignin NDF ratio. [b] Means in the same column with different superscripts differ (p<0.05).

According to the obtained results the hybrids ZP 427, ZP 648 and ZP 666 (Table 7) were superior to other investigated hybrids regarding IVDMD and, therefore, rated as very suitable for feed production. The hybrids ZP 427 and ZP 648 share one common parent, while the hybrids ZP 648 and ZP 666 are genetically related. Even though genetically similar to ZP 427, ZP 341 had statistically lower IVDMD; however, it was still rated as good for feed preparation due to sufficiently high IVDMD of the whole plant. The remaining two hybrids ZP 758 and ZP 802 were not suitable for the purpose regarding very low IVDMD. These hybrids have one common parent and the other parental components are very similar.

Considering the crucial effect and the great importance of observed parameters of maize yield and biomass quality on the estimation of hybrid fitness for silage, the intercorrelation between parameters was determined.

Table 8. Correlation coefficients between whole plant IVDMD and lignocellulose fibers content of maize hybrids[a]

	ADF	ADL	Hemi-cellulose	Cellulose	NFDD	LNDF^{-1}	Dry matter digestibility
NDF	0.94**	0.86**	0.95**	0.91**	-0.18	0.67*	-0.74**
ADF		0.83**	0.77**	0.99**	-0.14	0.67*	-0.66*
ADL			0.80**	0.76**	0.58*	0.95**	-0.93**
Hemi-cellulose				0.73**	-0.21	0.60*	-0.73**
Cellulose					-0.04	0.58*	-0.58*
NDFD						-0.73**	0.79**
LNDF^{-1}							-0.91**

[a] NDF-Neutral Detergent Fibres, ADF-Acid Detergent Fibres, ADL-Acid Detergent Lignin, NDFD-digestibility of Neutral Detergent Fibres, LNDF^{-1}-ADL-lignin NDF ratio. [b] Marked correlations are significant at p<0.05, N=6.

A highly significant negative correlation was observed between digestibility of the whole maize plant and NDF, ADL, hemicelluloses content and lignin to NDF ratio (r=-0.74, r=-0.93, r=-0.73, r=-0.91, respectively) and a significant negative correlation between the ADF and cellulose content and the IVDMD (r=-0.66 and -0.58, respectively). A very significant positive correlation was determined between NDF and ADF, ADL, hemicelluloses, cellulose, as well as among ADF and ADL, hemicelluloses and cellulose content. The NDFD was positively significantly correlated with ADL and ADL with hemicellulose and cellulose content (Table 8).

A very significant negative correlation was also found between L/NDF of the whole maize hybrid plants and NDFD and IVDMD (r=-0.73, r=-0.91, respectively). Based on the similar results, Riboulet et al. (2008) concluded that a maize hybrid, intended for a high value animal feed, should have a low content in ADL/NDF (ADL-lignin NDF ratio, %), the first factor negatively influencing whole plant digestibility. The results of the study performed by Seven et al. (2006) also indicated the strong relationship between nutritional composition and feed digestibility.

Understanding how maize plant cell wall constituents affect IVDMD is an important goal of future breeding research programs in order to improve forage utilization in animal feeding.

CONCLUSION

Our studies on technological (wet-milling) properties of the selected maize hybrids showed high recovery (>90%) and yield (>65%) of starch, as well as a low content of proteins (<0.3%). The hybrids with the increased starch content and a lower hectoliter mass and density, and a greater proportion of soft endosperm fraction had higher yields and recovery of starch in wet milling.

Kernel physical traits and chemical composition significantly varied among tested hybrids. This research confirms that the hardness (the combination of several physical quality traits) of maize kernel is very important criterion and clearly had significant association with chemical composition (starch, protein and cellulose contents) in maize kernel. Therefore, hardness parameters could be used to predict nutritional quality and utility value of maize kernels.

Based on the acquired results, it can be concluded that a maize hybrid, intended for a high value animal feed, should have a low content in ADL/NDF (ADL-lignin NDF ratio, %), because it negatively influences whole plant digestibility. Accordingly, a very significant positive correlation was found among all assayed lignocellulose fibers components.

The obtained results point out to the necessity of the preliminary assessment of newly released maize hybrids, and determination of parameters affecting its utilization value for different purposes to meet demand for higher yield and technological and nutritional properties. Our findings indicate that genetic differences embedded through decades of maize hybrids breeding programs can lead to providing farmers and industry with hybrids of good quality, desirable traits, and acceptable yield under variable agroecological conditions and management inputs. However, variability of quality parameters of the maize hybrids under different environmental conditions and soil, manure and pest control should be further investigated in future studies.

REFERENCES

(AOAC) Association of Official Analytical Chemists (1990). *Official Methods of Analysis*, Ed. K. Herlich, AOAC, Arlington, VA, pp. 70–84.

(AOAC) Association of Official Analytical Chemists (2000). *Official Methods of Analysis, 17th international edition - AOAC International*, Gaithersburg, MD., USA. Methods 923.03, 925.09, 930.15, 955.04, 960.39.

(ISO) International Organization for Standardization (1993). *Agricultural food products. De-termination of crude fibre. General method NF-V03-040 (status: certified standard ref. ISO 5498)*. Assn. Fr. De Normalisation, Paris.

(ISO) International Organization for Standardization (1997). *Determination of starch content-Ewers polarimetric method*. International Standard: ISO 10520.

Aufréré, J., (2006). *Prevision de la digestibilite des fourages et aliments concentres et composes des herbivores par une methode enzymatique pepsine-cellulase*. AQ 353, pp. 1-6. [*Prediction of the digestibility of forages and concentrated and compound feeds of herbivores by an enzymatic pepsin-cellulase method.*]

Bertoia, L. A., Aulicino M. B. (2014). Maize forage aptitude: Combining ability of inbred lines and stability of hybrids. *The Crop Journal* 2: 407-418.

Boisen, S., Hvelplund, T., Weisbjerg, M.R. (2000). Ideal amino acid profiles as a basis for feed protein evaluation. *Livestock Production Science*, 64: 239–251.

Brenner, E. A., Zein, I., Chen, Y., Andersen, J. R., Wenyel, G., Ouyunova, M., Eder, J., Darnhofer, B., Frei, U., Barriere, Y., Lubberstedt, T. (2010). Polymorphisms in O-methyltransferase genes are associated with stover cell wall digestibility in European maize (*Zea mays* L.). *BMC Plant Biology* 10, 27. doi:10.1186/1471-2229-10-27.

CIMMYT (2003). The development and promotion of quality protein maize in sub-Saharan Africa. *Progress report submitted to Nippon foundation.* CIMMYT, Mexico.

Coleman, C. E., Larkins, B. A. (1999). The pro-lamins of maize. In: *Seed proteins,* Eds. P. R. Shewry, R. Casey, Kluwer, Dordrecht, pp. 109–139.

Correa, C. E., Shaver, R. D., Pereira, M. N., Lauer, J. G., Kohnt, K. (2002.). Relationship between corn vitreousness and ruminal in situ starch degradability. *Journal of Dairy Science* 85: 3008-3012.

Dupont, J., White, P. J., Carpenter, M. P., Schaefer, E. J., Meydani, S. N., Elson, C. E., Woods, M., Gorbach, S. L. (1990). Food uses and health effects of corn oil. *Journal of the American College of Nutrition* 9: 438–470.

Eckhoff, S. R. (1999). *High-extractable starch corn: What is it? Wet Milling Notes*, Note No.17, February. Department of Agricultural Engineering, University of Illinois, USA.

Ferraretto, L. F., Shaver, R. D. (2015). Effects of whole-plant corn silage hybrid type on intake, digestion, ruminal fermentation, and lactation performance by dairy cows through a meta-analysis. *Journal of Dairy Science* 98: 2662–2675 http://dx.doi.org/ 10.3168/jds.2014-9045.

Fox, G., Manley, M. (2009). Hardness methods for testing maize kernels. *Journal of Agriculture and Food Chemistry* 57(13): 5647-5657. doi:10.1021/jf900623w

Frey, T. J., Coors, J. G., Shaver, R. D., Lauer, J. G., Eilert, D. T., Flannery, P. J. (2004). Selection for silage quality in the Wisconsin quality synthetic and related maize populations. *Crop Science* 44: 1200-1208.

Hegyi, Z., Zsubori, Z., Rácz, F., Halmos, G. (2009). Comparative analysis of silage maize hybrids based on agronomic traits and chemical quality. *Maydica* 54: 133-137.

Huang, S., Frizzi, A., Florida, C. A., Kruger, D. E., Luethy, M. H. (2006). High lysine and high tryptophan transgenic maize resulting from the reduction of both 19- and 22-kD alpha-zeins. *Plant Molecular Biology* 61: 525-535.

Jaeger, S. L., Luebbe, M. K., Macken, C. N., Erickson, G. E., Klopfenstein, T. J., Fithian, W. A., Jackson, D. S. (2006). Influence of corn hybrid traits on digestibility and the efficiency of gain in feedlot cattle. *Journal of Animal Science* 84: 1790–1800.

Johnson, L. M., Harrison, J. H., Davidson, D., Robutti, J. L., Swift, M., Mahanna, W. C., Shinners, K. (2002). Corn silage management. I Effects of hybrid, maturity and mechanical processing on chemical and physical characteristic. *Journal of Dairy Science* 85(4): 833-853.

Ketthaisong, D., Suriharn, B., Tangwongchai, R., Jane, J., Lertrat, K. (2015). Physicochemical and morphological properties of starch from fresh waxy corn kernels. *Journal of Food Science and Technology*, DOI 10.1007/s13197-015-1750-2.

Khan, N. A., Yu, P., Ali, M., Cone, J. W., Hendriks, W. H. (2015). Nutritive value of maize silage in relation to dairy cow performance and milk quality. *Journal of the Science of Food and Agriculture* 95(2): 238-252.

Krakowsky, M. D., Lee, M., Coors, J. G. (2006): Quantitative trait loci for cell wall components in recombinant inbred lines of maize (*Zea mays* L.) II: leaf sheath tissue. *Theoretical and Applied Genetics* 112: 717–726.

Kumar, S., Sarkar, A., Singh, R. P., Singh, R. (2018). Agro-environmental consequences of quality protein maize (QPM) hybrid development with special emphasis of soil nitrogen management. *Plant Archives* 18(1): 147-157.

Lauer, J. G., Coors, J., Shaver, R. (2001). What's coming down the pike in corn genetics? Value added corn silage - brown midrib, waxy, high-oil and others. *The 31st California Alfalfa and Forage Symposium*, Modesto, CA. Proceedings, p. 159-171.

Lorenz, A. J. (2009). *Characterization, inheritance, and covariation of maize (Zea mays L.) traits relevant to cellulosic biofuels production.* UMI, ProQuest LLC, Ann Arbor, MI, pp. 1-178.

Maize Annual Report (2018). maize.org/maize-annual-report-2018/.

McGrance, S. J., Cornell, H. J., Rix, C. J. (1998). A simple and rapid colorimetric method for the determination of amylose in starch products. *Starch* 50: 158-163.

Mertens, D. R. (1992). Critical conditions in determining detergent fibers. In: *Forage Analysis Workshop*, Proceedings, NFTA Forage Testing Assoc. Denver, CO. Natl. Forage Testing Assoc. Omaha, NE, pp. C1-C8.

Milasinovic, M., Radosavljevic, M., Dokic, Lj., Jakovljevic, J. (2007). Wet-milling properties of ZP maize hybrids. *Maydica* 52 (3): 289-292.

Milašinović-Šeremešić, M., Radosavljević, M., Srdić, J., Tomičić, Z., Đuragić, O. (2019). Physical traits and nutritional quality of selected Serbian maize genotypes differing in kernel hardness and colour. *Food and Feed Research* 46(1): 51-59.

Milašinović-Šeremešić, M., Radosavljević, M., Terzić, D., Nikolić, V. (2018). Maize processing and utilisation technology-achievements and prospects. *Journal on Processing and Energy in Agriculture* 22(3): 113-116.

Milašinović-Šeremešić, M., Radosavljević, M., Terzić, D., Nikolić, V. (2017). The utilisable value of the maize plant (biomass) for silage. *Journal on Processing and Energy in Agriculture* 21 (2): 86-90.

Moore, S. M., Stadler, K. J., Beitz, D. C., Stahl, C. H., Fithian, W. A., Bregendahl, K. (2008). The correlation of chemical and physical corn kernel traits with growth performance and carcass characteristics in pigs. *Journal of Animal Science* 86: 592–601.

Muehlbauer, G. J., Gengenbach, B. G., Somers, D. A., Donovan, C. M. (1994). Genetic and amino-acid analysis of two maize threonine over producing, lysine-in sensitive aspartate kinase mutants. *Theoretical and Applied Genetics* 89: 767-774.

Pomeranz, Y., Czuchjowska, Z., Martin, C. R., Lai, F. (1985). Determination of corn hardness by Stenvert hardness test. *Cereal Chemistry* 62: 108–110.

Radosavljevic, M., Bekric, V., Milasinovic, M., Pajic, Z., Filipovic, M., Todorovic, G. (2010). Genetic variability as background for the achievements and prospects of the maize utilisation development. *Genetika* 42 (1): 119-136.

Radosavljević, M., Milašinović Šeremešić, M., Terzić, D., Todorović, G., Pajić, Z., Filipović, M., Kaitović, Ž., Mladenović Drinić, S. (2012). Effects of hybrid on maize grain and plant carbohydrates. *Genetika* 44(3): 649-659.

Radosavljević, M., Terzić, D., Semenčenko, V., Milašinović Šeremešić, M., Pajić, Z., Mladenović Drinić, S., Todorović, G. (2015). Comparison of selected maize hybrids for feed production. *Journal on Processing and Energy in Agriculture* 19(1): 38-43.

Riboulet, C., Lefevre, B., Denoub, D., Barriere, Y. (2008). Genetic variation in maize wall for lignin content, lignin structure, p-hydroxycinnamic acid content, and digestibility in set of 19 lines at silage maturity. *Maydica* 53: 11-19.

Schwab, E. C., Shaver, R. D., Lauer, J. G., Coors, J. G. (2003). Estimating silage energy value and milk yield to rank corn hybrids. *Animal Feed Science and Technology* 109: 1-18.

Scott, M. P., Edwards, J. W., Bell, C. P., Schussler, J. R., Smith, J. S. (2006). Grain composition and amino acid content in maize cultivars representing 80 years of commercial maize varieties. *Maydica* 51, 417-423.

Semenčenko, V., Mojović, L., Đukić-Vuković, A., Radosavljević, M., Terzić, D., Milašinović-Šeremešić, M. (2013). Suitability of some selected maize hybrids from Serbia for the production of bioethanol and dried distillers' grains with solubles. *Journal of the Science of Food and Agriculture* 93(4): 811-818.

Semenčenko, V., Radosavljević, M., Terzić, D., Milašinović-Šeremešić, M., Mojović, Lj. (2014). Dried distillers' grains with solubles (DDGS) produced from different maize hybrids as animal feed. *Journal on Processing and Energy in Agriculture* 18(2): 80-83.

Šeremešić, S., Nikolić, I., Milošev, D., Živanov, M., Dolijanović, Ž., Vasiljević., M. (2016). The possibility of maize gluten application for weed control in maize and soybean. *Bulgarian Journal of Agricultural Science,* 22(1), 52-59.

Seven, P. T., Cerci, I. H. (2006). Relationships between nutrient composition and feed digestibility determined with enzyme and nylon bag (in situ) techniques in feed resources. *Bulgarian Journal of Veterinary Medicine* 9(2): 107-113.

Singh, S. K., Johnson, L. A., Pollak, L. M., Hurburgh, C. R. (2001). Compositional, physical, and wet-milling properties of accessions used in germplasm enhancement of maize project. *Cereal Chemistry* 78: 330-335.

Spackman, D. H., Stein, W. H., Moose, S. (1958). Automatic recording apparatus for use in the chromatography of amino acids. *Analytical Chemistry* 30: 1190–1206.

STATISTICA (Data Analysis Software System) (2018). v.13.3., Stat-Soft, Inc., USA (www.statsoft.com).

Szyszkowska, A., Sowinski, J., Wierzbicki, H. (2007). Changes in the chemical composition of maize cobs depending on the cultivar, effective temperature sum and farm type. *Acta Scientiarum Polonorum, Agricultura* 6(1): 13-22.

Terzić, D., Radosavljević, M., Milašinović-Šeremešić, M., Semenčenko, V., Todorović, G., Pajić, Z., Vančetović, J. (2012). Lignocellulose fibres and in vitro digestibility of ZP maize hybrids and their perental inbred lines. *Third International Scientific Symposium "Agrosym 2012"* Jahorina, November 15-17, 2012, Book of Abstracts, 51, Book of Proceedings, 209-214.

Thomas, E. D., Mandebvu, P., Ballard, C. S., Sniffen, C. J., Carter, M. P., Beck, J. (2001). Comparison of corn silage hybrids for yield, nutrient composition, in vitro digestibility, and milk yield by dairy cows. *Journal of Dairy Science* 84, 2217-2226.

Wang, L., Xu, C., Qu, M., Zhang, J. (2008). Kernel amino acid composition and protein content of introgression lines from *Zea mays* ssp. mexicana into cultivated maize. *Journal of Cereal Science* 48, 387-393.

Watson, S. A. (2003). Description, development, structure and composition of the corn kernel. In *Corn: Chemistry and Technology,* 2nd ed. Eds. J.

White, L. Johnson, American Association of Cereal Chemists, St. Paul, MN, pp. 69–101.

Zehr, B. E., Eckhoff, S. R., Nyquist, W. E., Keeling, P. L. (1996). Heritability of product yields from wet-milling of maize grain. *Crop Science* 36: 1159-1165.

Zehr, B. E., Eckhoff, S. R., Singh, S. K., Keeling, P. L. (1995). Comparison of wet-milling properties among maize inbred lines and their hybrids. *Cereal Chemistry* 72: 491-497.

Zilic, S., Milasinovic, M., Terzic, D., Barac, M., Ignjatovic-Micic, D. (2011). Grain characteristics and composition of maize specialty hybrids. *Spanish Journal of Agricultural Research* 9(1): 230-241.

In: *Zea mays* L.: Cultivation, and Uses
Editor: Sarah Dunn

ISBN: 978-1-53619-181-3
© 2021 Nova Science Publishers, Inc.

Chapter 3

THE EFFECT OF SALICYLIC ACID IN MAIZE BIOPRODUCTIVITY

C. J. Tucuch-Haas[1,*], G. Alcántar-González[2],
L. Trejo-Téllez[2], H. Volke-Haller[2], Y. Salinas-Moreno[3],
J. I. Tucuch-Haas[3], M. A. Dzib-Ek[4], S. Vergara-Yoisura[4]
and A. Larque-Saavedra[4]

[1]Tecnológico Nacional de México/ITS del Sur de Yucatán,
Mexico City, México
[2]Colegio de Postgraduados, Mexico City, México
[3]Instituto Nacional de Investigación Forestal Agrícola y Pecuario,
Mexico City, México
[4]Centro de Investigación Científica de Yucatán, Merida, México

---

[*] Corresponding Author's E-mail: cesar_5204@hotmail.com.

Abstract

Mesoamerican cultures are generally regarded as advanced societies that, among other contributions to humanity, are known to have domesticated cultivated plants as *Zea mays*. Maize is one of the staple foods of the Mexican population and the practice of nixtamalization of maize seeds before Spanish conquest in 1521, is fundamental in the preparation of dough for tortillas.

We have shown that applications of low concentrations of salicylic acid (SA) in plant seedling shoots or in evergreen trees significantly increase growth, development, and productivity. In order to assess the effect of spraying, low concentrations (SA) in maize seedling in development conduct experiments in growth rooms that have shown that 1 μm of SA significantly increased root length by 30.6% and 0.1 M of SA 24.7% compared to control. This concentration also significantly increased the total fresh biomass of seedlings.

In other experiments the results have shown that (SA) significantly increase the length, weight and dry weight of roots, stems, leaves and yield of this species, as well nitrogen (N), phosphorus (P) and potassium (K) levels in the different organs of plants at harvest time. Copper, zinc, manganese, iron, boron, calcium, and magnesium were also increased in most tissues by the effect of SA. It is proposed that the positive effect of SA of increasing root size promotes the absorption and accumulation of macro and micronutrients and contributes to seed production.

Keywords: salicilyc acid, foliar spray, root length, photosynthesis, phenols, farm yield

Introduction

Mesoamerican cultures are considered advanced societies that, among other contributions to humanity, are known to have domesticated cultivated plants such as Maize (*Zea mays*). Maize is one of the staple foods of Mexican population and the practice of nixtamalization of maize seeds, biotechnology prior to the Spanish conquest in 1521, is fundamental in the preparation of dough for tortillas.

The growing demand for grain, due to the population growth (Jurado et al., 2013) and climate changes, such as the increase in temperatures and the greatest precipitation variability (Gerald et al., 2009), has exceeded the offer, despite the increase in yields achieved based on research focused on genetic improvement (Santiago-Lopez et al., 2017) and the supply of chemical and organic fertilizers (Alvarado et al., 2018). That is why a change in the orientation of research has been urgently needed, betting on the use of growth regulators, as is these case with salicylates.

Followed by the observation of its role in opening and closing of stomatic, acetylsalicylic acid, *in plants of Phaseolus vulgaris* L. and *Commelina communis* L. (Larque-Saavedra, 1978 and 1979) in the 70's these compounds were proposed to regulate physiological processes, by our working group, as a potential strategy in agricultural production, triggered a series of investigations aimed at highlighting their contribution in the bioproductivity of plants.

Several works have been conducted in a wide variety of plant species angiosperms and gymnosperms, under controlled and field conditions, to demonstrate that, in particular, salicylic acid (SA), has great potential to positively impact radical development, flowering and productivity; when it is supplied exogenously (Larqué-Saavedra and Martín-Mex, 2007; Martín-Mex et al., 2013). In addition, it suggests that concentrations of the order of one micromole or below it, are sufficient to trigger such responses as demonstrated in hybrid *Petunia* plants in which it induced greater number of flowers and precocity in flowering (Martín-Méx et al., 2010), vegetables such as *Capsicum chinense* J. (Martin-Mex et al., 2004 and 2005) and *Cucumis sativus* (Martin-Mex et al., 2013) where 23 and 33% respectively, fruit yield was reported; and in the cultivation of *Carica papaya* L. in which it favored 21.9%.

In particular, the impact on root system (length and differentiation of secondary roots is reported) as a result of the supply of low concentrations of SA (Gutiérrez-Coronado et al., 1998; Echeverría-Machado et al., 2007; Larqué-Saavedra et al., 2010; Tucuch et al., 2015), is considered a key effect on the bioproductivity of plants, since it allows a greater absorption of nutrients, hypothesis that was evidenced in bioassays developed in habanero chili plants (Tucuch-Haas et al., 2017).

IMPACT OF THE SA ON THE RADICAL SYSTEM

The root plays a crucial role in the development and production of grain in maize cultivation because it is the main organ responsible for supplying water and nutrients to plants, for proper functioning in all vital activities. Therefore, increasing root mass, for a higher volume of soil exploration by root, is one of the key factors to induce the crop to a higher yield. Tests aimed at demonstrating the sensitivity of maize root to SA indicated a significant response, favoring up to 30.6% the length with 1µm and 24.7% volume with 0.1 µM (Tucuch et al., 2016; Figure 1).

The sensitivity of root system of maize plants, in conjunction with wheat showed a 34% increase in the length and 30% the weight of root was reported, sprayed with 1 µM of SA to the canopy of seedlings (Tucuch et al., 2015), demonstrate the ability of SA to induce favorable radicular development in grasses; (San Miguel et al., 2003), as well in *Chrysanthemum morifolium* (Villanueva-Couoh et al., 2009) and *Lycopersicum esculentum* (Larque-Saavedra et al., 2010) which pointed a sensibility range of 0.01 to 1.0 µM. In this regard Echeverría-Machado et al., (2007) and Shakirova et al., (2003), suggest that this response is possible due to the induction of SA, in the increase level of cell division of apical meristem, the size of cofia and lateral roots.

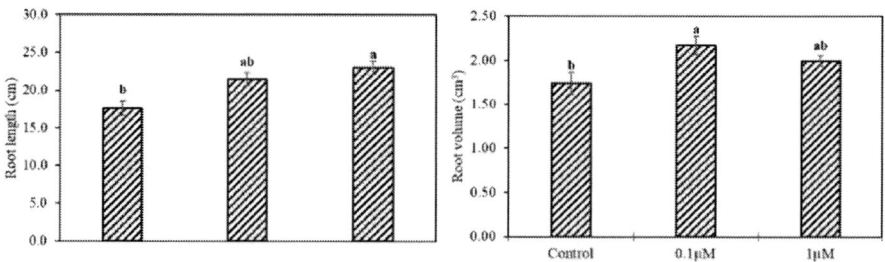

Figure 1. Effect of salicylic acid on the length and volume of corn seedling root (var. Xmejen-nal). Each block is the average of 7 individuals±S E. Stockings with different letters mean they are significantly different.

EFFECT OF SA ON AERIAL BIOMASS

The research carried out to study the benefit of SA in biomass, accuses, up to a 48% increase in total dry biomass, when 1 µM is sprayed in the canopy at the seedlings with the presence of a leaf, for five consecutive days and evaluated ten days after last application. Under this same method of application, but evaluated, at harvest stage (end of experiment) up to 68% is reported, in addition significant increases in stem height and diameter (Tucuch-Haas et al., 2017a; Table 1). Similar effects have been reflected by other authors, when plants are subjected to salinity stress (Keshavarz et al., 2019) or drought (Ghazi, 2017). This behavior is attributed to the SA ability to induce a greater number of leaves and their foliar expansion (Farouk et al., 2018), which consequently increases a higher photosynthetic rate.

Table 1. Effect of application of salicylic acid to the canopy of maize plants (var. Xmejen-nal), in plant height, stem diameter and aerial dry biomass

Experiment	Treatment	PH (cm)	SD (mm)	TDB (g plant-1)
1	Control	213.50 b	16.03 b	218.4 b
	0.1 M	224.50 ab	17.69 b	279.0 b
	1.0 M	244.00 a	22.65 a	368.50 a
2	Control	206.22 a	19.98 a	229.03 b
	0.1 M	205.77 a	20.04 a	242.65 b
	1.0 M	218.40 a	20.28 a	347.49 a

PH: plant height; SD: stem diameter; TDB.: total dry biomass. Means with the same letter in each column for each experiment, are not significantly different (Tukey, 0.05).

SA ON THE NUTRITIONAL STATUS OF MAIZE

Due to the relevant role that nutrition plays in the development of maize, given its participation in many physiological and metabolic processes, two experiments were proposed to corroborate the participation of SA in nutritional plant status. The data collected, unlike calcium in which there was no consistency in response, in two experiments results showed significant benefits in content of nitrogen (N), phosphorus (P), potassium (K), magnesium (Mg) and sulfur (S) in foliar biomass (75, 85, 67, 46 and 62% respectively) and grain (59, 129, 116, 120 and 86% respectively); as well as iron (Fe), copper (Cu), zinc (Zn), manganese (Mn), boron (B) and sodium (Na) in these same organs (Tucuch et al., 2017a; Table 2 and 3). This response trend and those found in the cultivation of habanero chili (Tucuch et al., 2017), support the hypothesis that suggests that the main mechanism of AS to increase bioproductivity is the absorption of nutrition, due to the sensitivity of this compound to promote greater root development.

SA Involvement in Photosynthesis

In conjunction with radical system, photosynthesis is another of determining factors in the bioproductivity of maize, so that many of the efforts to increase yields are based on the selection or creation of varieties with a great photosynthetic capacity. Compounds of a phenolic nature, with the potential to influence the opening and closing stomates and other physiological processes, such as salicylic acid, have been proposed.

The participation of the SA in photosynthesis, respiration and stomatic conductance, in the cultivation of maize, was demonstrated by an experiment conducted maize in the open field with a local variety of maize sprayed with 0.1 and 1.0 µM. Patterns of photosynthesis behavior, stomatic conductance and plant respiration are shown in Figure 2. The values indicate that photosynthesis was significantly affected at 15:00 h by 28% with 1.0 µM and with 13% with 0.1 µM of SA (Figure 2). They also play a crucial role in controlling photosynthesis and its components, as it is able to maintain the CO_2 levels of leaves, in hours with highest radiation, even at the expense of stomatic closure, and increase it in the hours with least radiation. This effect on photosynthesis could be explained as a result of an increase in Rubisco enzyme activity by the action of SA as reported by Khodary (2004).

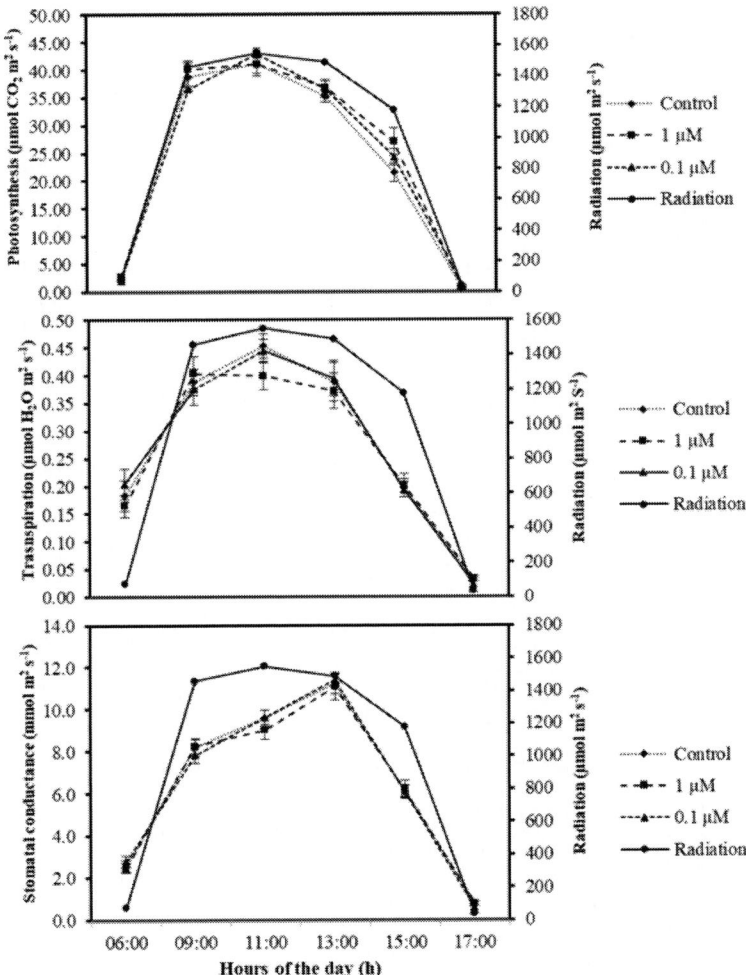

Figure 2. Effect of salicylic acid on photosynthesis, stomatal conductance, and respiration of xmejen-nal variety corn culture. Each data is the average of 10 plants±SE.

Table 2. Effect of SA on macro and micronutrient content on corn plant var. Xmejen-nal, assessed 140 days after transplantation

		Macronutrients (g plant^{-1})						Micronutrients (g plant^{-1})					
		N	P	K	Ca	Mg	S	Fe	Cu	Zn	Mn	B	ln
Exp. 1	Control	1.41 c	0.13 b	0.64 c	1.77 c	0.70 c	0.19 c	0.13 c	0.001 b	0.007 c	0.009 c	0.003 c	0.06 c
	0.1 M	1.94 b	0.13 b	1.02 b	3.57 a	1.00 a	0.24 b	0.44 b	0.001 b	0.010 b	0.018 b	0.005 b	0.07 b
	1 M	2.79 a	0.26 a	1.34 a	2.88 b	1.05 a	0.33 a	1.22 a	0.003 a	0.017 a	0.063 a	0.006 a	0.12 a
Exp.2	Control	2.01 b	0.24 b	1.04 b	12.66 a	1.01 b	0.27 b	1.72 a	0.002 b	0.008 b	0.03 a	0.004 b	0.17 b
	0.1 M	2.25 b	0.23 b	0.84 c	3.81 c	0.83 c	0.23 c	0.63 c	0.001 c	0.006 c	0.01 b	0.003 c	0.15 c
	1 M	3.10 a	0.42 a	1.31 a	8.91 b	1.44 a	0.41 a	1.37 b	0.004 a	0.012 a	0.03 a	0.008 a	0.28 a

Table 3. Effect of AS on macro and micronutrient content on corn plant grain, var. Xmejen-nal, assessed 140 days after transplantation

		Macronutrients (mg cob^{-1})						Micronutrients (mg cob^{-1})					
		N	P	K	Ca	Mg	S	Fe	Cu	Zn	Mn	B	ln
Exp. 1	Control	1020.6 c	145.4 c	62.99 b	9.77 b	82.63 c	66.03 c	2.99 b	0.107 a	1.30 b	0.23 b	0.69 a	5.49 b
	0.1 M	1831.8 b	314.7 b	140.5 a	21.88 a	170.69 b	123.76 b	5.08 a	0.158 a	2.48 a	0.41 a	1.12 a	10.70 a
	1 M	2096.5 a	393.4 a	146.6 a	19.03 a	200.44 a	136.94 a	3.62 ab	0.145 a	2.66 a	0.46 a	0.68 a	9.32 a
Exp.2	Control	2200.9 b	233.3 b	84.5 c	11.0 b	117.8 b	108.7 b	2.4 b	0.09 c	2.2 a	0.42 b	0.29 b	5.05 b
	0.1 M	1657.4 c	301.1 b	118.9 b	14.0 ab	155.8 b	114.2 b	3.4 ab	0.17 b	2.7 a	0.36 b	0.39 b	7.32 b
	1 M	2548.4 a	420.60	158.4 a	21.6 a	234.9 a	180.9 a	4.4 a	0.26 a	3.1 a	0.56 a	0.57 a	10.29 a

EFFECT OF SALICYLIC ACID ON YIELD

Table 4. Effect of salicylic acid on the cob and grain of maize cultivar (Xmejen-nal) sprayed at seedling stage in two independent experiments

	Treatment	Cw (g)	Ng (#)	Rg (g cob-1)
Exp. 1	Control	79.4 b	240.0 b	57.5 b
	0.1 M	141.8 ab	362.5 ab	109.2 ab
	1 M	164.5 a	374.5 a	120.2 a
Exp. 2	Control	106.9 b	134.5 b	62. 6 b
	0.1 M	134.8 ab	192.8 b	90.9 ab
	1 M	185.9 a	349.3 a	140. 1 a

CW: cob weight; NG: number of grains; RG: Grain performance per cob. Means with the same letter in each column for each experiment are not significantly different (Tukey, 0.05).

The challenge of today's agriculture is to increase crop yields with sustainable techniques, in addition to improving production, they are friendly to the environment y and economically accessible to the producer. For maize, one of technologies assessed for this purpose, is the foliar supply of salicylic acid, to the culture canopy, in seedling stage, with significant benefits. In this regard Tucuch-Haas et al., (2017a) conducting two open-field experiments in order to know the impact on production, reporting an increase of 60% the number of grain, 90% the weight of a cob and 100% the yield of the grain above control, in both experiments, in maize plants cv. Xmejen-nal, treated with 1µm of SA (Table 4). The clarity in the response of this compound, can also be observed in species such as wheat and rice where yield benefits of 17 and 60% respectively have been reported (López et al., 1998; Tavares et al., 2014).

The sensitivity of performance to SA is subject to a variety of factors, environmental assumptions and dosages, so that, drought conditions, the supply of 1 mM of AS to plants with 10 to 12 exposed leaves, increases 59% the yield of the grain (Zamaninejad et al., 2013); and the application of 100 mg L^{-1}, to plants of the variety SC131, at 40 and 60 days after planting induces an increase of 27% (Farouk et al., 2018). When subjected to 25% water deficit and 200 mg L^{-1} is sprayed at the canopy, plants of the

Giza10 variety, at 25 days of age, it favors 40%grain yield (Ghazi, 2017), however, with 1 mM favors a15% the yield in the cv Research 106 (Jasim et al., 2017); while with 300 mg L^{-1} the profit was 26% (Ahmad et al., 2018).

EFFECT OF SA ON GRAIN PHENOLS

Phenols in plants play a significant role in pigmentation processes, growth, reproduction, adaptation to biotic and abiotic stress conditions and many other functions (Farah and Marino, 2006; Lattazio et al., 2006). Particularly in maize, some phenolic compounds have been linked to coloration (Salinas-Moreno et al., 2013), hardness (Cabrera-Soto et al., 2009) and tolerance to pathogens such as *Fusarium* (Bakan et al., 2003) and *Sitophilus zeamays* (García et al., 2003). Salicylic acid, when supplied exogenously, to maize plants, in seedling stage, accelerates the accumulation processes of total phenols, increasing up to 16% the concentration with 1 µM and 4% with 0.1 µM (Tucuch-Haas et al., 2017b; Figure 3), which could explain one of the reasons, of its participation in the tolerance to pathogens reported in the literature.

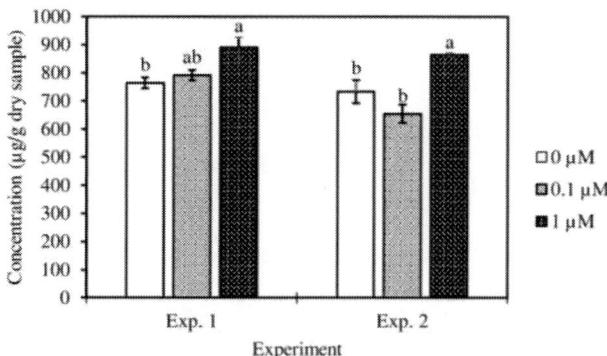

Figure 3. Effect of SA on the concentration of phenols on grain sprayed in the canopy of seedling-stage maize plants. Each block is the average of 4 repiclates±S E. Means with the same letter are not significantly different.

CONCLUSION

The large amount of beneficial responses by effect of SA, in maize, such as modification of the root system, the dry biomass, photosynthesis and phenol content in the grain, reflecting an yield's increase, suggest that this compound as a new alternative to mitigate the production deficit of the maize crop. Concentration of 1 µM of SA to the growing canopy, in, seedling stage (first leaf with ligula), for five days, is sufficient to trigger such responses.

REFERENCES

Ahmad, H., Khan, I., Liaqat, W., Faheem, J. M. and Ahmadzai, D. M. (2018). Effect of Salicylic Acid on Yield and Yield Components of Maize under Reduced Irrigation. *Int J Environ Sci Nat Res*, 9 (3): 76-80.

Alvarado, T. R., Aceves, R. E., Guerrero, R. J. D, Olvera, H. J., Bustamante, G. A., Vargas, L. S. and Hernández, J. H. (2018). Response of maize genotypes (*Zea mays* L.) to different fertilizers sources in the Valley of Puebla. *Terra Latinoamericana*, 36(1): 49-59.

Bakan, B., Bily, A. C., Melcion, D., Cahagnier, B., Regnault, Philogene B. J. R. and D. Richard-Molard (2003). Possible role of plant phenolics in the production of trichothecenes by *Fusarium graminearum* strains on different fractions of maize kernels. *J. Agric. Food Chem.*, 51: 2826-2831.

Cabrera-Soto, M. L., Salinas-Moreno, Y., Velázquez-Cardelas, G. A. and Espinosa, E. T. (2009). Content of soluble and insoluble phenols in the structures of corn grain and their relationship with physical properties. *Agrociencia*. 43 (8): 827-839.

Echeverría-Machado, I., Escobedo-GM, R. M. and Larqué-Saavedra, A. (2007). Responses of transformed *Catharantus roseus* roots to femtomolar concentrations of salicylic acid. *Plant Physiol Biochem*, 45:501-507.

Farah, A. and Marino, C. D. (2006) Phenolic compounds in coffee. *Braz. J. Plant Physiol,* 18: 23-36.

Farouk, S., Arafa, S. A. and Nassar, R. M. A. R. (2018). Improving drought tolerance in corn (*Zea mays* L.) by foliar application with salicylic acid. *International Journal of Environment*, 7 (3): 104-123.

García, S. L., Burt, A. J., Serratos, J. A., Díaz, D. M. P, Arnason, J. T. and Bergvinson D. J. (2003) Natural defenses in maize grain to attack by *Sitophilus zeamais* (Motsch, Coleoptera: Curculionidae): mechanisms and bases of resistance. *Rev. Educ. Bioq.*, 22(3): 138-145.

Gerald, C. Nelson, Rosegrant, M. W., Koo, J., Robertson, R., Sulser, T., Zhu, T., Ringler, C., Msangi, S., Palazzo, A., Batka, M., Magalhaes, M., Valmonte-Santos, R., Ewing, M. and Lee D. (2009). *Climate Change: The Impact on Agriculture and the Costs of Adaptation.* International Food Policy Research Institute Washington, D.C.

Ghazi, A. D. (2017). Impact of Drought Stress on Maize (*Zea mays*) Plant in Presence or Absence of Salicylic Acid Spraying. *J. Soil Sci. and Agric. Eng.*, 8 (6): 223-229.

Gutiérrez-Coronado, M. A., Trejo-López, C. and Larqué-Saavedra, A. (1998). Effects of salicylic acid on the growth of roots and shoots in soybean. *Plant Physiol. Biochem.,* 36: 563-565.

Jasim, A. H., Hasson, K. M and Rashid, M. H. (2017). The effect of salicylic acid and phosphorus spraying on maize (*Zea mays* L.) yield under conditions of incomplete irrigation. *Annals of West University of Timişoara, ser. Biology.* 20 (1):21-30.

Jurado, A. J., Gutiérrez, A. H, Callejas, J. N. and Ortega, M. F. I. (2013). Economic Situation of Corn Production in Terms of Irrigation in the State of Chihuahua. *Revista Mexicana de Agronegocios*, 33:504-512.

Keshavarz, Y., Alizadeh, O., Sharfzade, S., Zare, M. and Bazrafshan, F. (2019). Investigating the Importance of Salicylic Acid and Mycorrhiza in Reducing the Unfavorable Effects of Stresses on Maize. *J. Mater. Environ. Sci.,* 10 (12): 1401-1412.

Khodary, S. E. A. (2004). Effect of salicylic acid on the growth, photosynthesis and carbohydrate metabolism in salt stressed maize palnts. *Int. J. Agri. Biol.*, 6: 5-8.

Larque-Saavedra, A. (1978). The Antiranspirant Effect of Acetylsalcylic Acid on *Phaseolus vulgaris*. *Physiologia Plantarum*, 43(2):126-128.

Larqué-Saavedra, A. (1979). Stomatal Closure in Response to Acetylsalicylic Acid Treatment. *Zeitschrift Für Pflanzenphysiologie*, 93(4):371-375.

Larqué-Saavedra, A. and Martín-Mex, R. (2007). Effects of salicylic acid on the bioproductivity of the plants. In Hayat S. y A. Ahmad (Eds), *Salicilylic acid, a plant hormone*. Springer publishers, Dortdrech, The Netherlands.

Larqué-Saavedra, A., Martin-Mex, R., Nexticapan-Garcez, A., Vergara-Yoisura, S. and Gutierrez-Rendón, M. (2010). Effect of salicilic acid on the growth of tomato (*Lycopersicon esculentum* Mill.) seedlings. *Rev. Chapingo Ser. Horticultura*, 16 (3):183-187.

Lattanzio, V., Lattanzio, V. M. T. and Cardinali A. (2006) Role of phenolics in the resistance mechanisms of plants against fungal pathogens and insects. *Phytochemistry: Adv. Res.*, 661: 23-67.

López, T. R., Camacho R.V. and Gutiérrez C. M. A. (1998). Use of Salicylic Acid Sprays on Wheat to Increase Yield in Three Wheat Varieties. *Terra Latinoamericana*, 16 (1): 43-48.

Martín-Mex, R., A. Nexticapan-Garcéz and A. Larqué Saavedra. (2013). Potential benefits of salicylic acid in food production. In Hayat S., A. Ahmad and M. N. Alyemeni (Eds). *Salicylic acid*. Springer publishers, Dortdrech, The Netherlands.

Martín-Mex, R., López-Gutiérrez, R., Medina-Arceo, J., Cruz-Campos, J., Nexticapan-Garces, A., González-Rodríguez, F. and Larqué-Saavedra, A. (2004). Increase in the productivity of habanero pepper (Capsicum chinense Jacq.) Due to salicylic acid sprays. *First World Convention of Chili*. León, Guanajuato, México. p. 326.

Martín-Mex, R., Nexticapan-Garcés, A., Vega-Merino, L., Baak-Polanco, A. y Larqué-Saavedra, A. (2005). Effect of salicylic acid on the flowering and productivity of habanero pepper (*Capsicum chinense*

Jacq.). *Second World Convention of Chile chili.* Zacatecas, Zacatecas, México. p. 325-326.

Martín-Mex, R., Vergara-Yoisura, S., Nexticapán-Garcés, A. y Larqué-Saavedra, A. (2010). Application of low concentrations of salicilyc acid increases the number of flowers in *Petunia hibrida. Agrociencia,* 44: 773-778.

Salinas-Moreno, Y., García, C. S., Estrada, E. C. and Vida, V. A. M. (2013). Content and type variability of anthocyanins in blue/purple colored grains of Mexican maize populations. *Rev. Fitotec. Mex.,* 36:285-294.

San-Miguel, R., Gutiérrez, M. and Larqué-Saavedra, A. (2003). Salicylic acid increases the biomass accumulation of *Pinus patula. South J Appl For*, 27:52-54.

Santiago-López, N., García-Zavala, J. J., Mejía-Contreras, A., Espinoza-Banda, A., Santiago-López, U., Esquivel-Esquivel, G. and Molina-Galán, J. D. (2017). Grain yield of Tuxpeño corn populations adapted to High Valleys de México. *Revista Mexicana de Ciencias Agrícolas*, 8 (1): 145-156. [*Mexican Journal of Agricultural Sciences*]

Shakirova, F. M., Sakhabutdinova, A. R., Bezrukova, M. V., Fatkhutdinova, R. A. and Fatkhutdinova, D. R. (2003). Changes in the hormonal status of wheat seedlings induced by salicylic acid and salinity. *Plant Sci,* 164: 317-322.

Tavares, L. C., Araújo, R. C., De Olivera, S., Pich, B. A., and Amaral V. F. (2014). Treatment of rice sedes with salicylic acid: seed physiological quality and yield. *J. Seed Sci.,* 36: 356-356.

Tucuch, H., C. J., Alcántar, G. G., and Larqué, S. A. (2015). Effect of Salicylic Acid on Root Growth and Total Biomass of Wheat Seedlings. *Terra Latinoamericana*, 33: 63-68.

Tucuch-Haas, C. J., Alcántar-González, G., Volke-Haller, V. H., SalinasMoreno, Y., Trejo-Téllez, L. I. and Larqué-Saavedra, A. (2016) Effect of salicylic acid on growth root maize seedlings. *Revista Mexicana de Ciencias Agrícolas*, 7 (3): 709-716. [*Mexican Journal of Agricultural Sciences*]

Tucuch-Haas, C. J., Pérez-Balam, J. V., Díaz-Magaña, K. B., Castillo-Chuc, J. M., Dzib-Ek, M. D., Alcántar-González, G., Vergara-Yoisura, S., and Larqué-Saavedra A. (2017). Role of salicylic acid in the control of general plant growth, development, and productivity. In R. Nazar et al., (eds.), *Salicylic Acid: A Multifaceted Hormone.* Springer Nature Singapore Pte Ltd.

Tucuch-Haas, C., Alcántar-González, G., Salinas-Moreno, Y., Trejo-Téllez, L. I.; Volke-Haller, V. H. and Larqué-Saavedra, A. (2017b). Leaf spraying of salicylic acid increases the concentration of phenols in maize grain. *Revista Fitotecnia Mexicana*, 40 (2): 235-238. [*Mexican Fitotecnia Magazine*]

Tucuch-Haas, C., Alcántar-González, G., Trejo-Téllez, L. I, Volke-Haller, H., Salinas-Moreno, Y., & Larqué-Saavedra, A. (2017a). Effect of salicylic acid on growth, nutritional status, and performance of maize (*Zea mays*). *Agrociencia*, 51 (7): 771-781.

Tucuch-Haas, C., Alcántar-González, G., Volke-Haller, H., SalinasMoreno, Y., Trejo-Téllez, L. and Larqué-Saavedra, A. (2015). Photosynthesis, transpiration, stomatal conductance and chlorophyll content in a maya landrace of maize treated with salicylic acid. *Wulfenia J.,* 22:375-381.

Villanueva-Couoh, E., Alcántar-González, G., Sánchez-García, P., Soria-Fregoso, M. and Larque-Saavedra, A. (2009). Effect of salicilic acid and dimethyl sulphoxide in the flowering of [*Chrysanthemum morifolium* (Ramat) Kitamura] in Yucatan. *Rev. Chapingo Ser. Horticultura*, 15:25-31.

Zamaninejad, M., Khorasani S. K., Moeini M. J. y Heidarian A. R. (2013). Effect of salicylic on morphological characteristics, yield and yield components of corn (*Zea mays* L.) under drought condition. *Euro. J. Exp. Bio.*, 3: 153-161.

INDEX

A

acid, vii, viii, x, 1, 2, 3, 4, 54, 55, 57, 65, 67, 78, 79, 84, 85, 86, 87, 88, 89, 91, 92, 94, 95, 96, 97
amino acid, 6, 54, 58, 65, 66, 69, 75, 79, 80
animal husbandry, 5
animal nutrition, v, ix, 37, 51, 52, 53, 54, 56
antibiotic, viii, 2, 4
antioxidant, viii, 2, 4

B

bacteria, vii, viii, 2, 4, 5, 6, 7, 8, 9, 12, 13, 14, 15, 16, 17, 18, 19, 23, 24, 25, 26, 27, 28, 29, 30, 31, 32, 33, 34, 35, 50
bacterial cells, 6, 13, 28
bacterial colonies, 8, 17
bacterial strains, 25
bacterium, 26, 27
base, 20, 21, 22, 23, 65
benefits, 88, 91, 95
bioavailability, 5
biomass, x, 3, 53, 57, 70, 72, 78, 84, 87, 88, 93, 96
biosynthesis, 5
biotechnology, 24, 48, 84
breeding, x, 52, 55, 59, 74
Burkholderia, viii, 2, 6, 22, 23, 24, 25, 28, 29, 30, 32, 34, 35, 44, 45, 47, 49

C

calcium, viii, x, 2, 3, 4, 5, 84, 88
carbohydrate, ix, 51, 54, 59, 64, 95
carbohydrate metabolism, 95
carbohydrates, 3, 54, 57, 70, 79
cellulose, 54, 57, 67, 68, 70, 71, 73, 74
chemical, vii, ix, x, 52, 53, 54, 56, 58, 60, 62, 64, 65, 67, 68, 70, 74, 76, 77, 78, 80, 85
chemical composition, vii, ix, 36, 52, 53, 54, 55, 56, 60, 61, 62, 64, 67, 70, 74, 80
chromatography, 58, 80
cluster analysis, 21, 59, 66
commercial, ix, 52, 54, 79

composition, vii, ix, 7, 11, 52, 53, 54, 55, 56, 59, 60, 61, 62, 64, 65, 66, 67, 69, 70, 73, 74, 79, 80, 81
compounds, viii, 2, 3, 4, 5, 24, 85, 94
conductance, 88, 89, 97
constituents, x, 52, 74
correlation, ix, 52, 58, 60, 62, 63, 67, 68, 73, 78
correlation analysis, 60, 67
correlation coefficient, 58, 60, 62, 67, 68
crop, 3, 5, 31, 53, 65, 70, 86, 91, 93
cultivation, 17, 85, 86, 88
culture, 8, 9, 12, 13, 16, 18, 29, 89, 91

D

degradation rate, 34
denaturation, 11, 20, 31
digestibility, viii, ix, 2, 3, 5, 6, 52, 53, 57, 70, 71, 72, 73, 74, 76, 77, 79, 80
digestion, 53, 76
digestive enzymes, viii, 2
distilled water, 7, 8, 13, 15, 33
dough, vii, x, 84
drought, 59, 87, 91, 94, 97
dry matter, ix, 52, 54, 57, 64, 70, 71

E

electric current, 20
electric field, 20
electrophoresis, 11, 19, 20, 21
elongation, 20
endophyte, 2, 38, 39, 41, 49
endosperm, vii, ix, 52, 54, 55, 58, 61, 62, 63, 64, 67, 68, 74
energy, vii, ix, 1, 3, 31, 52, 54, 55, 64, 70, 79
environment, 4, 24, 28, 65, 69, 91
environmental conditions, 56, 65, 75
environmental factors, ix, 52

enzyme, viii, 2, 5, 6, 14, 24, 26, 29, 31, 32, 34, 35, 64, 80, 89
enzyme inhibitors, 64
ethanol, 3, 8, 10, 55
exposure, 14, 31
extraction, 9, 19, 56

F

farm yield, 84
fibers, x, 52, 70, 71, 73, 74, 78
flotation, 61, 62, 64
foliar spray, 84
food, 3, 35, 55, 59, 65, 75, 95
food production, 95
formation, 14, 20, 29

G

genetic background, 54, 66, 71
genetic diversity, 55
genetic engineering, 6, 55
genetics, ix, 52, 56, 67, 77
genus, 23, 24, 25, 26, 27
growth, x, 4, 6, 8, 13, 18, 26, 27, 28, 29, 35, 55, 78, 84, 85, 92, 94, 95, 96, 97
growth hormone, 26

H

hardness, ix, 52, 55, 58, 67, 69, 74, 78, 92
hemicellulose, ix, 52, 70, 71, 73
hybrid, ix, 52, 53, 56, 59, 60, 65, 71, 72, 73, 74, 76, 77, 79, 85
hydrolysis, 4, 6, 17, 33, 34, 57
hypothesis, 85, 88

I

in vitro, 11, 20, 30, 33, 34, 35, 80

in vivo, 33
incubation time, 13, 28
incubator, 8, 13, 15, 16, 29
individuals, 86
induction, 5, 18, 24, 86
industry, ix, x, 52, 53, 59, 69, 74
interdependence, 62, 68
ions, viii, 2, 3, 31
iron, viii, x, 2, 3, 84, 88

K

kernel, ix, 52, 53, 54, 55, 58, 59, 61, 62, 63, 64, 65, 67, 68, 74, 78, 80

L

lactation, 53, 76
lignin, ix, 52, 54, 70, 71, 72, 73, 74, 79
lysine, 6, 25, 54, 65, 68, 76, 78

M

magnesium, viii, x, 2, 3, 84, 88
maize, v, vii, ix, x, 1, 2, 3, 4, 5, 7, 8, 9, 12, 14, 15, 16, 17, 18, 19, 23, 24, 27, 30, 31, 32, 33, 34, 35, 51, 52, 53, 54, 55, 56, 57, 58, 59, 60, 62, 64, 65, 66, 67, 68, 69, 70, 71, 72, 73, 74, 75, 76, 77, 78, 79, 80, 81, 83, 84, 86, 87, 88, 91, 92, 93, 94, 95, 96, 97
maize biomass, 53, 70
maize hybrids, vii, ix, 52, 53, 56, 57, 59, 60, 62, 64, 65, 66, 70, 71, 72, 73, 74, 76, 78, 79, 80
maize kernel, ix, 52, 54, 55, 58, 59, 64, 66, 67, 68, 74, 76, 93
management, 53, 65, 74, 77
manganese, viii, x, 2, 3, 84, 88
matter, iv, 57, 58, 70, 71, 72, 73
measurement, 13, 15, 31, 32, 33
media, 7, 8, 9, 12, 16, 17, 18, 24, 29
metabolism, 3, 4, 5, 25
micronutrients, x, 84
microorganisms, 19, 23
molecules, 20, 21, 31
multiple regression, 63
multiple regression analysis, 63

N

nitrogen, viii, x, 2, 4, 26, 35, 58, 65, 77, 84, 88
nitrogen fixation, 26
nucleotide sequence, 11, 12, 23, 27
nutrients, 5, 24, 54, 70, 85, 86
nutrition, ix, x, 3, 52, 53, 54, 56, 70, 87
nutritional quality, vii, ix, 52, 56, 69, 74, 78
nutritional status, 97

O

oil, 55, 56, 59, 64, 68, 76, 77
organic compounds, 25
organs, x, 7, 8, 84, 88

P

pathogens, viii, 2, 35, 92, 95
pH, 7, 12, 14, 25, 30, 31, 32, 35
phenolic compounds, 92
phenols, 84, 92, 93, 97
phenylalanine, 6, 65
phosphate, 4, 5, 13, 14, 18, 21, 24, 25, 26, 29, 30, 31
phosphates, viii, 2, 4, 35
phosphorus, viii, x, 2, 3, 4, 84, 88, 94
photosynthesis, 84, 88, 89, 93, 95, 97
phylogenetic tree, 12
physical properties, 93

physical traits, vii, ix, 52, 53, 56, 64, 74, 78
phytase, v, vii, viii, 1, 2, 4, 5, 6, 7, 8, 9, 12, 13, 14, 15, 16, 17, 18, 19, 24, 25, 26, 27, 28, 29, 30, 31, 32, 33, 34, 35, 36, 37, 38, 39, 40, 41, 42, 43, 45, 46, 47, 49, 50
phytic acid, vii, 1, 2, 3, 4, 36, 39
pigmentation, 9, 19, 92
plant growth, viii, 2, 4, 25, 97
plants, vii, viii, x, 2, 4, 6, 24, 25, 26, 27, 33, 35, 57, 70, 71, 73, 84, 85, 86, 87, 89, 91, 92, 95
polysaccharides, 70
population, vii, x, 84
population growth, 84
positive correlation, ix, x, 52, 62, 73, 74
poultry, vii, viii, 1, 2, 3, 4, 5, 30, 31, 35
preparation, iv, vii, x, 7, 16, 72, 84
problem-solving, viii, 2
production costs, 5
proteins, viii, 2, 3, 5, 54, 57, 59, 65, 69, 74, 76
purity, ix, 52, 59, 63

R

recovery, ix, 52, 59, 60, 61, 62, 63, 74
regression, 14, 16, 29, 62
regression analysis, 62
regression equation, 14, 16, 29
response, viii, ix, 2, 52, 58, 64, 67, 68, 86, 88, 91
room temperature, 13, 14, 15, 16, 33, 50
root length, x, 84
root system, 85, 86, 93
roots, viii, x, 2, 8, 17, 30, 35, 84, 85, 93, 94

S

salicilyc acid, 84, 96
scientific understanding, 56
seedlings, x, 84, 86, 87, 95, 96

sequencing, 12, 19, 21
sodium hydroxide, 21
solution, 11, 13, 14, 15, 20, 29, 57
species, x, 8, 9, 12, 16, 19, 23, 24, 25, 26, 27, 84, 85, 91
starch, ix, 3, 52, 54, 55, 56, 58, 59, 60, 61, 62, 63, 64, 67, 68, 69, 74, 75, 76, 77, 78
sterile, 7, 8, 10, 15, 29, 33
storage, viii, 2, 3, 4, 8, 69
structural changes, 31
structure, 19, 54, 70, 79, 80
substrate, viii, 2, 5, 6, 14, 24, 31, 34

T

taxonomy, 24, 25, 26
temperature, 12, 14, 25, 30, 31, 32, 35, 60, 80
traits, vii, ix, 52, 53, 55, 56, 59, 64, 65, 67, 68, 69, 74, 76, 77, 78
treatment, 6, 14, 15, 30, 33
tryptophan, 54, 65, 76

V

variations, 6, 14, 30, 31, 32
varieties, 66, 79, 88

W

water, 4, 8, 14, 15, 30, 64, 86, 91
water absorption, 64
wet milling, v, x, 51, 52, 53, 56, 59, 60, 62, 74, 76
whole plant, ix, 52, 53, 54, 57, 70, 71, 72, 73, 74

Y

yield, viii, ix, x, 2, 4, 6, 35, 52, 53, 56, 59, 60, 61, 62, 63, 66, 70, 72, 74, 79, 80, 84, 85, 86, 91, 93, 94, 96, 97

Z

Zea mays L., v, vii, 1, 2, 3, 16, 28, 29, 35, 49, 76, 77, 93, 94, 97